MUCK GRIME & SWEAT

IGNITING A LIFE IN FIRE

*To Adam
Stay Safe
Darran*

DARRAN GOUGH

EU Conformity Declaration

This product complies with the following safety regulations and standards to ensure consumer safety and product quality: Regulation (EU) 2023/988 of the European Parliament and of the Council on General Product Safety (GPSR): The Consumer Product Safety Improvement Act (CPSIA), Section 101. The Californian Safe drinking water and toxic enforcement act. (Proposition 65) EN71-Part 1: Mechanical and Physical Properties EN71-Part 2: Flammability EN71-Part 3 Migration of certain elements.

Published by Snorticle Stories
Produced and Manufactured by Softwood Books
EU Responsible person: Maddy Glenn
Office 2, Wharfside House, Prentice Road, Stowmarket, Suffolk, IP14 1RD
www.softwoodbooks.com
hello@softwoodbooks.com

EU Rep:
Authorised Rep Compliance Ltd., Ground Floor, 71 Lower Baggot Street, Dublin, D02 P593, Ireland
www.arccompliance.com
info@arccompliance.com

Text © Darran Gough, 2025
Illustrations © Amy Cremmen and Roger Dighton
Cover phographer: Richard Pinches
Cover makeup: Kim Nicholls

Paperback ISBN: 978-1-0683709-0-8
Hardback ISBN: 978-1-0683709-1-5

ACKNOWLEDGEMENTS

I would like to thank the following people for their assistance and patience with the production of this book.

Andy Howland.

My maths teacher at Gillott's school back in the 70s. We have remained friends ever since and he was the first person to start reading my drafts and helping me with the text from a non-service point of view.

Tom Carroll and Mike Smyth. These gents were some of my Chief and Deputy Chief Fire Officers at Oxfordshire Fire Service. They have helped me with some of the technical terms and making sure I don't slip into too many abbreviations.

Maurice Johnson. My first Chief Fire Officer for writing the introduction.

Fiona Hammans. A former colleague from my school governor days. A respected head and someone I trust to tell me if bits are not right. She once told me that she always wanted me to sit on disciplinarily panels because "I don't take any shit". A complement that I also return to her.

Family wise, my cousins Nina and Margaret. Nina from a remote, (she now lives in Australia) but family objective. Margaret for an absolutely amazing workload to proofread the book for me.

My friend from photography college Richard Pinches, for creating a marvellous cover photo for the book. It brought back many happy memories of the 80s in the studios and darkrooms in Reading.

Kim Nicholls for the fantastic makeup job. I really have come out of fire looking like that.

Graham Gegg for letting me use the black and white photo he took of me as a very young inexperienced fireman after just taking a lung of smoke in, seriously frowned upon today.

Roger Dighton and Amy Cremmen my tattoo artists for all the wonderful illustrations.

All at Softwood Books for steering me through the publication of this my first book.

My stepdaughter Claire for writing my bio for the book.

My wife Sheila for putting up with me for over 22 years now. Overdue her long service medal, the fire service is 20 years. Which by the way she will get in a couple of years as she is also in the fire service.

And finally, my dear but departed mum, Joyce Gough. For all she put up with, on her own, from a stroppy teenager after loosing his dad, and entering a dangerous career without thinking how it would affect her. For all the support she gave me, especially showing me how to iron my uniform shirts!

Darran Gough
C8 Billesley Fire Station West Midlands Fire Service January 2025

REVIEWS

"Darran's depth of knowledge and experience in each area of the fire and rescue service role is quite unique. I cannot think of another serving whole time firefighter who has also served as an on call firefighter and received 999 calls as a fire control officer. It is that depth of understanding of each vital component of the modern fire service that makes Darran's book such a compelling read."

– Mike Smyth, Deputy Chief Officer Oxfordshire Fire and Rescue Service, retired

"This is a superb read, written with great enthusiasm. There is much drama and a fair bit of humour."

– Andy Howland

"A great read. Darran has produced a lovely story of the journey through his inimitable career."

– Tom Carrol. Chief Fire Officer Oxfordshire Fire Service Retired

"As the book developed chapter by chapter, I really wanted to know what was going to happen next - it was gripping."

"A great tale of achieving an ambition and the reality of the firefighter's role."

– Fiona Hammans

"From midnight callouts and racing to save lives, this fireman's story takes us on a journey of a lifetime given to the fire service. An engaging book that keeps you reading and leaves you uplifted."

– Nina Hedger

FOREWORD
Much is owed by many to just a few

No disrespect is intended by amending the context of a perennial saying that relates to those who fought in the 39/45 war. There is another group of public servants who were then, and still are now, "in the firing line".

They are referred to "Retained" firefighters - so called because, although otherwise employed in their local area, a relatively small "retaining fee" is paid to secure their services and protect local communities. Day and night they guard the interests of residents and property in most of the suburban and rural areas of the UK. Their skill and commitment mirrors that of their colleagues in the towns and cities who perform their duties as a fulltime job.

Apart from routine passive fire and accident prevention work the nation's Fire and Rescue services respond to and aid those in danger or distress from almost any natural or human crisis. Dramatic fire scenes are mirrored by a variety of accident situations, flooding and other disasters big and small. So often it is the gentle hand of someone in a yellow helmet that helps a human (or animal) in distress – perhaps not able to resolve all the problems but a friendly face in time of need.

This book sets out to describe one person's experience from a unique point of view. The author has served in the three primary branches of the service i.e. in the retained section, as a career firefighter, and also as a member of Control staff responsible for receiving emergency calls and despatching resources to deal with them.

This is his story.

– *Oxfordshire Chief Fire Officer (Retired) Maurice Johnson*

The author would be content that terminology used in the book is appropriate to the era the work was set in and may not be in line with modern-day language norms in the context of today, which is wholesomely evidence and scrapers.

Opinions in the book are the author's and not necessarily those of the organisation he has worked for.

Where some identities have been changed others have been kept the same with their permission.

CHAPTER 1
Childhood Memories

The mournful, eerie wailing of the siren drifted into Woolworths just as I was about to choose some 'pick and mix'. Suddenly the thought of sweets was replaced with an urgency for my mum to take me in the pushchair out of the shop and to the crossroads in the middle of the town.

As we turned the corner, I could see men running in the same direction, up the marketplace and past the Town Hall to disappear out of sight. The siren faded and the town was silent. Then in the distance came the familiar two-tone sound of the Henley-on-Thames retained fire crew. I am reliably informed by my mother that I would shout out at the top of my voice:

"Nee Naw mummy Nee Naw!"

It would swing around the front of the Town Hall and disappear up Gravel Hill. Mum would then try and turn the pushchair back to where we had come from, but I wasn't having any of it. I knew all too well that Henley had two fire engines and I wanted to see both.

So, sure enough, a couple of minutes later the second distant "Nee Nor" sound would start, and the gleaming red and silver sister engine would swing into view from behind the Town Hall and follow the water trail from the first engine up the hill and away. With a beaming smile on my face and a sense of satisfaction, we could now return to the more mundane task of choosing my 'pick and mix'.

It was a school night, and I was fast asleep in my bedroom. The door opened and in came my mother silhouetted by the landing light. She came over and gently shook me, "Darran get up and come and look at this".

My father was the lock keeper at Shiplake Lock and because of this our house was on an island in the middle of the Thames. As my mother drew back the curtains I could see on the other side of the river, behind the weeping willows, a huge fire. One of the holiday cottages was burning fiercely.

The flames were jumping high into the sky and the shadows of the willows were dancing on the weir stream making thousands of patterns, all different and strangely graceful as they were mixed with the sparks and burning embers landing in the stream. Some embers were extinguished as soon as they landed but the bigger ones remained alight and drifted along with the current until they faded out of sight.

Even though my bedroom window was closed, I remember the noise that the fire made. The constant cracking, popping and banging as the wooden bungalow slowly became engulfed by the flames.

The cottage, even though it was close to me just on the other side of the river, was very remote from the nearest village of Wargrave, and the track to the cottage was very long and rough.

It seemed an age before I saw the first reflections in the smoke of the fire engine's blue flashing lights in the distance.

Eventually the outline of a fireman came into view in front of the rapidly disintegrating home. I could hear him shouting out orders and waving his arms, giving directions to his crew somewhere in the darkness.

Another noise now joined the cacophony. It was a high pitch whirring mechanical sound. It was the pump on the engine starting up.

Suddenly the light from the fire caught the water from a fireman's hose as it arched up into the night sky like the tail of a comet, bright red and orange. I sat on the windowsill in the darkened room with my mother, transfixed for ages by what I was watching. Occasionally I would see the outlines of more firemen as they carefully moved around the collapsing house.

"Come on Darran I think you've seen enough now, back to bed".

I reluctantly slid down from the windowsill and got back into bed. My mother kissed me goodnight and went back downstairs. I lay there watching the flickering patterns on my curtains and listening hard for anything I could hear. Eventually I drifted off to sleep again.

The next morning, I was woken up by our unique alarm clock. Our Dachshund Rema would come running up the stairs and jump on my bed licking my face until I gave in and got up. I went straight to the window, but as I pulled the curtains back, the view that greeted me was totally different from the one I had witnessed a few hours previously.

There was no noise, no bright dancing flames, no shouting, for behind the willows was nothing but a huge pile of smouldering timbers, with wisps of smoke lazily lifting from the charred remains of the once quaint riverside cottage.

I could now make out the firemen carefully picking their way over the wreckage with their hoses and pitch forks. Their Yellow leggings now black and their faces matching their smoke smeared yellow helmets.

"Breakfast is ready" came the call from downstairs, so I turned away and plodded down to the kitchen.

After breakfast and with my satchel packed, I went out the back door to start my walk to school. I opened the door, and a smell invaded my nostrils that I would recognise and identify all my working life. It was the smell of the remains of the house. I had smelt burning wood before. I frequently helped my dad with the numerous bonfires he made to get rid of the driftwood that would clog up the lock gates, but this was different. There was more to it than just burning wood. It had the aroma of the furnishings, the contents, personal belongings and such like that were no more. I was still only at Primary School and didn't realise the enormity of what I had witnessed. All I knew was I had one hell of a story for Miss Newman and my class that morning!

My next encounter with the world of the fireman was on a Primary School trip to the Oxfordshire County Fire Service training school at Didcot fire station. We were all excitedly ushered through the muster bay with all the neatly lined up fire kit on pegs and I saw something that seemed a bit strange to me at the time. All the firemen's trousers were stood up around their boots!

We went past the glistening fireman's pole and into the appliance bay. The smells that met us were that of a place of work. Aromas of oil, petrol and smoke mixed with that of disinfectant and polish made a cocktail of images and wonderment as we entered this magical place.

In front of us were two enormous bright red and silver gleaming fire engines, completely quiet and apparently sleeping, but poised to spring into action when the siren sounded. The things that caught my eye were the two huge red wheels hanging off the back of one of them. They were about 6 feet high and would not look out of place behind a shire horse pulling a cart load of hay. The fireman explained that these were on the wheeled escape ladder. We were shown around the engines and all the equipment they carried. We climbed inside the cab and sat on the bench seat. Behind us were the breathing apparatus sets with their masks and air cylinders hanging from the racks on the wall.

I recognised some of the equipment, as it was in my Ladybird book "People who Help Us, The Fire Service" The silver T shaped standpipe for getting the water out of the hydrants.

In one locker (these are what the equipment is kept in on the side of the engine) was a big engine that slid out on a tray. This was called the lightweight portable pump or LPP. This was used to get water out of rivers and ponds when they couldn't get the fire engine near enough.

There were bags of rope, they called them lines, but I was fascinated by the locker with the axes, hammers and the hydraulic cutting gear. This bright red jigsaw puzzle consisted of a hand pump, cutters and spreaders as well as a ram. It was all connected by small black hoses and was used to cut people out of crashed cars.

The final locker was full of canvas hoses all rolled neatly in rows with brass couplings with two lugs either side in the middle and over the top the open snake like mouth with some drips of water staining the hose from the last time it was used.

At back of the engine, between the two huge wheels of the ladder, was the heart of this giant, the pump. This again was gleaming red with 4 brass outlets for the hose, behind it sat the water tank and more control dials and levers.

We were then taken out to the training ground and told to stand very still on the drill line. Little did I know that in about 13 years' time I was to have an encounter on that very line with a man that would haunt me for many years to come.

From around the corner of the building came the two leviathans awoken from their slumber and roaring with energy. Lights flashing and two tones sounding they swung around in front of us and as then screeched to a halt, the doors were flung open, and the firemen leaped out from the cabs.

Four of the men ran to the rear of one of the engines and pulled off the giant wheeled escape ladder and with two of them spinning the enormous wheels, they raced it toward the drill tower. As they approached, one of them at the rear started winding a handle and the ladder started extending upwards towards the top floor window. As it clattered into the opening the fourth member of the crew was already running up the ladder. He reached the window and after stamping on the sill he disappeared inside.

While this apparently effortless ladder pitch was happening, the other firemen were scurrying around the yard setting into the hydrant and running out the canvas hose, shouting out orders for water. As the pumps were turned on, I recognised that growling noise I had heard the night of the cottage fire. The hoses jigged and slithered around the floor as the pressure built up inside them. This was followed by the crackling of the water erupting at the end, as the fireman holding the kicking hose staggered back a couple of feet until he got a firm footing.

More and more hoses started erupting with water. The noise from the pumps grew louder and louder until eventually there was water pouring into every window of the tower.

Out of the top window reappeared the fireman who had run up the ladder, but this time he was moving very slowly as over his shoulders was a man-sized dummy. He very carefully grabbed hold of the top of the ladder and swung his right foot onto one of the rounds (non-fire people call them rungs) Then with his full weight on that foot he pulled himself back out of the window and onto the ladder. He steadied himself for a moment then began slowly climbing down the ladder, one round at a time.

As he reached the bottom a fireman in a white helmet shouted out at the top of his voice, "Knock off and make up" at the same time holding his arms out and dropping them firmly to his side. This command was shouted back to him by the crew and the pumps ceased roaring and the water dropped down as the hoses collapsed like a soufflé that has been taken out of the oven too early.

While the firemen were putting away the equipment, the one in the white helmet came over to us and showed us his uniform. He explained that his white helmet meant he was the Station Officer, and he was in charge. He showed us his thick leather belt with his fireman's axe hanging off it. He let us try on his helmet and I was surprised at how light it was. He explained that was because it was made of cork. He had big black leather boots poking out from under his yellow trousers. I asked him why the trousers were standing up on the boots in the station? He explained that the leggings, as they were called, were kept that way so they could put their feet straight into their boots and pull them up quickly.

When we had finished asking our questions all the crew had gathered around us and we all in chorus said, "Thank you" and we all clapped.

I am sad that when I did eventually join the fire service the majestic, wheeled escape ladders had finally been taken out of service a year before and replaced with what was called a 13.5 meter aluminium ladder. It still took four people to manoeuvre it, but it wasn't the same and I never have had the chance to use or climb the wheeled escape.

As time went by and I moved up to Secondary School and thoughts of the fire service were replaced with homework, part time jobs including the milk round, the butchers round and numerous different paper rounds in the village. Occasionally my future career would reappear in my life. One summer's afternoon I was eating my tea in the kitchen by the bay window overlooking the lock, when a commotion began outside. I went out and downstream from the lock there was a pall of smoke and people running towards it. In the middle of the lock cut (the part of the river leading up to the lock) was a boat a Freeman launch with smoke coming from the engine. I saw my dad running back to the lock office to call the fire service as I ran down onto the point. This was a long thin strip of land that the boats moored up on waiting to go through the lock and the boat was in the cut between the point and the far bank of private waterfront houses. There were people on board in and it was drifting towards the far bank. The boat got nearer and nearer to the bank and some new mooring decking. It finally got to within jumping distance and the people on board made the leap with the mooring ropes. They tied it to the bank and stood well back.

It is a strange but comforting feeling that a minute or so later the faint wailing of the Henley fire station siren, which called the men in, drifted on the wind, 3 miles away to the waiting crowd.

By this time the fire had taken hold and was spreading from the stern cockpit to the cabin. The homeowner had come running from his house down his long garden and was dragging his garden hose. When he turned it on instead of trying to put the fire in the boat out, he aimed it at his brand-new wooden deck mooring. The language he was aiming at the very distressed boat owners was very colourful and somewhat out of place as they had escaped with their lives and all he was concerned with was his mooring. To this day it never ceases to amaze me the priorities people put on possessions rather than lives.

A Freeman motor launch is made of fibreglass and as such burns, or rather melts, very quickly and while we were waiting for the fire service to arrive the gas bottles on board started to explode. These made the garden hose owner dive for cover and then reluctantly admit defeat and watch his decking join in the spectacle.

After what seemed like ages but in fact was probably only 10 minutes the familiar 'nee naw' sound came over the hill and down the lane. Next the firemen appeared in the garden opposite me with a lightweight portable pump like the one I had seen at Didcot all those years ago. They connected some large silver hose with a wicker basket on the end and threw it into the river. Next, they ran out a red hose and finally the pump was started with two or three violent yanks of the pull cord. They started putting water on to what was now a burning hull, only a couple of feet above the water line. It was put out in about 5 minutes, but the job was far from over. The cylinders that hadn't exploded still had to be dealt with. Using a long drag fork, they were carefully lifted from the wreck and lowered into the river to cool down. Excitement over, I wandered back to the house and finished my tea.

It's 1981 and I have finished Secondary School and while waiting for my O level results, I start a holiday job in a boat yard downstream from the lock. I had the perk of being able to use one of their boats to and from work every day. My job was to clean and ready the Thames slipper launches for day-trippers. I also had the lovely job of pumping out the toilets on the hire craft that call in on their holidays. These and other various jobs were made bearable by hearing the stories from Alan. Alan only worked in the yard every other 4 days, this was because he was a full-time fireman, and these were his days off. During our lunch breaks he would regale me with the shouts he had been on the previous shift, and hence my fire service interest was reignited.

One August afternoon, I moored the boat I had borrowed to get home from the boat yard after work, just below the lock and as I wandered up to the lock a friend on one of the permanent moorings said there's a letter waiting for you. A letter I thought so what?............MY EXAM RESULTS! I had completely forgotten about them. I ran the length of the lock and up the steps into the kitchen. Mum was holding the envelope out to me. I ripped it open and looked at it in shock. I'd passed all 7 subjects with B's and C's. I went straight out to show dad who was at the top gate, I got a bit of a smile and a well done, but that was all. My dad was of the old school, he was nearly 60 and, in his day, men didn't show their emotions especially not in public. I found out years later that he had said to my mum that he was proud of me and that I'd done really well, but he just couldn't say that to me. The results meant I could start at King James College in Henley studying A level English and Physics, AO Maths but more importantly A Level Photography.

January 3rd.1982, my world turned upside down. We had just had the whole family over for Christmas. I was 16, the youngest of three boys, and there was a 17 year gap between me and my nearest brother, Steve and another 3 to my eldest brother Roger. They both had families, and everybody had been staying in the lock house, so it had been a bit cramped, but it was the best Christmas I had ever known. Now in the New Year they had all gone home to Wales and the South Coast. That winter had been particularly wet, and the river was in flood. The water was right over the camping island our house was on, and the weirs were fully open. The last bridge upstream from us was Sonning, and before the holidays it had been undergoing some repair work. Unfortunately, the rising water levels had lifted the scaffolding planks off and they were floating down towards us. Shiplake weir was the first obstacle in their path and the planks became jammed across the weir. I was sitting in the living room as dad in full yellow oilskins went to go out the door to clear the weir. As he went to go out Tom and Jerry came on the TV, I called out to him as we both loved them. He stood leaning in the doorway watching, and when it finished, he turned and went out. That was the last time I ever saw my dad.

My friend came around and we went into the back room where I had a huge model railway layout, and we were playing with it when mum came in looking a bit frightened and said "Darran, dad's gone in the river"

I thought he had just slipped in the bank and would need a hand pulling him up, but as I ran out, I met the relief lock keeper who had been with dad, and he shouted to me that he had gone in the weir stream.

As I ran up the island, I remember thinking "I'm only wearing my slippers "and I was running through at least a foot of water. I made it around to the boat house, which faced onto the weir stream side of the island and was shouting at the top of my voice "DAD!" but all I got back was the echo. When a river is in flood, sound is amplified, and my cries were just rebounding off the water back at me.

I ran back down stream past the house and onto the point. By this time a small crowd, that had been out for a New Years walk, had gathered and I had to barge past them. As I did, I fell through a small hedge on the water's edge and nearly went straight into the weir stream myself, but someone grabbed me and pulled me back. I was beginning to realise what was happening and because the stream was running so fast, I ran with Stephen down the foot path through the posh house gardens, where the boat fire had been, downstream from the weir, looking for my dad. After about half an hour the realisation that there was very little we could do from the bank, we finally went back to the house. Someone had called an ambulance whose crew were now in the house trying to make my mum sit down, which was probably the worst thing they could do. She had no idea where I was and I think she feared the worst that I had gone in as well, which to be honest if I had seen dad I would have, without thinking and she would have probably lost me too. The reality was sinking in that we had lost dad. The only phone at the lock was the payphone in the lock office. Mum gave me a bag of 2p coins, and I went to phone my brothers. While I was doing this a good friend of the family came over the bridge just for a cuppa and walked right into this nightmare. This was the first time I broke down and he got a very wet shoulder. I will be for ever grateful to Roger Lightfoot for doing and saying very little but in that way doing a hell of a lot for me that day.

I got hold of my eldest brother Roger and he came straight back from the south coast. My middle brother was an outward-bound instructor and was out in the Welsh mountains when I called but when he got back to the centre, he called, and I broke the news to him as well. He came back overnight and as he arrived too early to wake us, he spent a couple of hours having a cuppa, in a large tug caller the "Churn" which was moored up at the lock and had been sent down from Reading to help in the search for dad when the light was better. That night me and Roger, made mum sleep with some tablets, but we couldn't sleep and spent most of the night playing cribbage, dad's favourite card game.

My whole life went on hold. All the local community went on the search for my dad. The local canoe club made numerous searches including the boat yards when the water level dropped enough to take boats out. It took 5 weeks to find him, and it was a canoe team from the Commandos who were practicing for the Devizes to Westminster Canoe race, who came across him just above Marsh, the next lock downstream. I remember seeing the policeman walking over the bridge towards the lock and before he got anywhere near the house mum said quietly "They've found him". It was February 7th, 1982, my 17th birthday. It was the best birthday present I could have had. Birthdays meant a lot to my dad, and he came back for mine.

This event was to have the most fundamental effect on my life, as you would expect. I became the man of the house looking after my mum at the age of 17. Not only that, but we were also going to be homeless. The lock house is tied with the job. The Thames Water Authority, as it was then, were fantastic. To put pressure on the local council to re-house us quickly, they served an eviction order on us. However, we were warned that this was coming and that they had no intention whatsoever of carrying it out. We could stay there for as long as needed. We saw some houses, but none were suitable and were miles away from our friends, who we needed so much. The local community started a fund for mum and I. This together with an insurance policy meant we could actually buy a house and in April that year we left the lock for the last time and moved into Henley. This may not seem much on first sight, but it moved me into a town that had a fire station, and more importantly I was living less than 5 minutes away.

I regularly visit the lock with my daughter Abigail, who never knew my dad. When she was little, she loved it when the lock keeper would let her work "Poppy's lock", the name that Roger and Steve's kids called dad. We also got a memorial bench installed and I and Abi take great pride in maintaining it, sanding and painting it every couple of years. We always, duty shift pattern permitting, go down on the 3rd of January each year and place some flowers at the weir side for dad.

I wish she had met him.

CHAPTER 2
"I Could Do That Job"

At the time I hadn't realised the impact the loss of dad had on my studies. At the end of my first year at King James's I had sat my mock exams three times. Eventually, I just put my name on the papers and sat there for two hours doing nothing. I was very fortunate that the Principal obviously had a far better idea what was happening to me than I did myself. Much against all the rules I was allowed to change my courses halfway through. I dropped A level Physics completely, A level English for O level English literature, the AO maths for O level art, but I was able to keep the A level photography.

About this time, I got a sort of cheer you up present of a new bicycle. It was a road bike, bright red and silver. I was longing to try it out, so the first chance came on a lovely sunny summer's evening, and I went for a ride out through Rotherfield Greys on a circular route of 4 or 5 miles, enough to shake out some cobwebs. I turned past The Maltster's pub and on towards the War Memorial. I was really getting into a rhythm. I turned onto a small track at the bottom of the triangle of the Memorial and headed back onto the Henley road, but as I came over the brow of the hill to head down through an overarching row of trees, down to Greys Court I saw a man jumping out of his car which was parked on the side of the road. Smoke and flames were beginning to erupt from under the bonnet. I shouted to him that I would go and call the fire brigade for him. I spun my bike around and peddled back as fast as I could to the house I had just passed. I dropped my bike on their drive and ran to the door and started banging on the door and ringing the bell for all I was worth. When a startled man opened the door, I blurted out "There's a car on fire down the road, I need to use your phone, where is it?" He beckoned me to the handset, and I dialled 999. I heard the operator say

"Emergency what service do you require?"

"Fire" I said somewhat breathlessly.

I was asked the phone number I was calling from, but it wasn't on the phone and the owner called it out to me. I heard the operator pass the number onto the Fire Control and I was then asked what the problem was?

"There's a car on fire on the Henley to….um" where was I? "Rotherfield Greys" the house owner shouted to me.

"Yes, yes Rotherfield Greys. It's on the steep hill coming up from the entrance to Greys Court," I added having now got my bearings again.

"Is there anyone in the vehicle?" the operator enquired.

"I saw the driver getting out, but I don't know about anyone else" I replied. I was asked a couple more questions and then was told the fire engine was on its way.

This was my first encounter with Fire Control, a place I would spend nearly 21 years working in. I put the receiver down and thanked the man for letting me use his phone. Now the car was around a dark blind bend, and I could see some cars going around it and not being able to stop. I decided to go to the top of the triangle of roads and divert the traffic down through the village and onto the back road to Henley, the one I had just cycled along. I was quite pleased with myself when every car I waved and indicated to, turned down the road. Occasionally one would stop and ask me what was going on and I would tell them. Then in the distance came a sports car, easily breaking the speed limit. I frantically waved at it to slow down and change course, but it came flying past me and disappeared over the brow of the hill. Then there was the sound of screaming brakes. I was expecting it to be followed by a thud, but nothing. A minute or so later the sports car came reversing shamefully, slowly back over the brow and back to the junction where I stood. I glared at the now embarrassed driver, a young man with, what must have been his girlfriend in the passenger seat. I let rip with a flow of verbal abuse about not looking at what I was trying to tell him, and he was lucky he wasn't killed. I don't think it worked as he just put his foot to the floor and left a smell of burning rubber as he sped off.

In the distance I could now hear the familiar two tones and along with them came a Panda car from the opposite direction. The policeman asked me what I was doing and when I told him he asked me to stay there and gave me a reflective waistcoat to wear. He then disappeared over the hill to see what was happening. After about half an hour, cars started to come past me from the direction of the fire and the policeman followed shortly after. He thanked me for my help and said I could go now. I pointed out to him that my new bike didn't have any lights on it yet, and I wasn't planning to stay out after dark, but was now going to get caught out. He told me to be careful on my way home, and if I was stopped, he gave me his card and said to say, he said it was ok.

During this time of turmoil, I met someone and we were friends for a couple of years or so, but more importantly she introduced me to her brother Graham. Graham was about 5 years older than me and single. He was a dispensing optician who just so happened to live in a flat that was next door to the Henley Fire Station. He loved his technology and was somewhat religious, playing the church organ every Sunday. Initially it was our mutual love of railways that cemented our friendship, but this was soon to move onto the fire service. Graham was friends with some of the fireman at the station, and he was also very interested in the newly developing world of video photography. The more I hung around with him the more I got to know the firemen. If I was at his flat when the bells went off in the station, we would run down the stairs and get into his little red VW Golf and follow the engines when they turned out. When we got to the fire or whatever they were going to, we would, with their permission, take some photographs of them dealing with the incident. We would develop the photo's ourselves and give them to the firemen for their scrapbooks. Some were very grateful and made no secret that they had one. Others would say thank you but that was all.

In 1983 I graduated from King James's with my A level in photography and passes in the English Lit and art. I still had no real idea of what I wanted to do for a career, but photography was taking up more and more of my time, so I applied and was accepted into the Berkshire College of Art and Design in Reading to do a BETEC Diploma in Photography. I was now learning and working in the photographic world and realised that the photos we were taking of the fire crew could earn us a few extra quid and help me with the college expenses. We decided to put our amateur scrapbook fillers onto a more professional footing. It always used to bug me that I would miss some of the calls if Graham couldn't get hold of me. Bear in mind this was 1983 and mobile phones were thin on the ground or at least like a brick to carry and only affordable to yuppies. What we did do was to rent a pager unit from BT. Little did I know but this was the beginning of having my life controlled by pagers for the next 30 or so years! I would carry it and if the crews turned out, Graham would page me, and I would drive down to his house to pick him up and away we would go. Unfortunately, it would take me 5 minutes or so to get to him and by then the fire engines would be long gone. If it was only a one pump shout there would be some crew left on station who would quietly mumble an address to us, as they weren't officially supposed to tell us where they had gone.

If both pumps had been mobilised and there was no one left on station we were in a bit of a pickle. However, there was a solution.

In those days when a call came in, the Fire Control would operate the fireman's pocket alerters or pagers, the days of the wailing siren were now long gone. The first man into the station would answer the telephone and write down on a pad the type of incident and the address. The paper had two carbon copies one for the second pump and one for the file. The small shelf that the pad and telephone were kept on was right by a window, and they would leave the pad where we could read it. So, while I was driving down to pick Graham up, he would wait until both engines had gone and then pop around to the window to read the pad.

This system used to work well until one cold snowy February night. It must have been about 2 in the morning when the pager woke me up. I got dressed quickly, (more about getting dressed later.) I grabbed my camera bag and headed outside. Mum's car was covered in snow, and I was swearing because I had forgotten to cover the windscreen. I got the door open and after a struggle with the manual choke, got the engine going. I then jumped out and with my un-gloved hands cleared the snow from the windows. The drive down to Graham's was treacherous with my wheels locking and the car sliding at the slightest touch of the brakes. I finally made it to the side of the station to find it empty, a two-pump shout. Graham was stood there looking somewhat despondent to say the least. Whoever had written the message down had let the cover of the pad fall back over the third copy, so the only piece of writing we could see didn't make any sense.

Graham said "Sorry to waste your time but there's nothing we can do, might as well go back to bed."

I was about to agree with him when an idea hit me. It was very early in the morning and the snow was still falling. There weren't many cars on the road except for idiots like us, AND FIRE ENGINES! All we had to do was to follow the tyre tracks of the engines on the road. It was worth a go and anyway, as I was out of bed and getting bloody cold, as the heater in my car still hadn't warmed up, there was no point going back to bed.

"Let's give it a go," we agreed. So, in the pitch black and light snow we set off both peering through the tiny clear patches in the still misted up windscreen at the 4 sets of tyre tracks. Down through the town and out towards Shiplake. Every now and then we would lose the tracks where the trees or bushes had shielded the road from the snow. We were grateful though that the tanks on the engines were full as every time they braked or turned a corner the water would slosh down the overflow pipe and leave a beautiful arrow shape in the road. Onward we went and eventually after about 15 minutes ended up in a little village called Binfield Heath. Here again we lost them, so I pulled over and opened my window. In all the freezing air I could smell the faint aroma of smoke and hear the familiar growling of the pumps. In the distance and reflected in the still gently falling snow we saw an orange glow mixed with flashes of blue, we were close. We moved off again towards the lights and eventually turned up an unmade track, and there they were alongside a house with a workshop blazing away and made more dramatic by the reflections in the still falling snowflakes.

I parked up a suitable distance away and we carefully walked up to them on the roadway made even more slippery by the now freezing water from the hoses. It was obvious we could not get any closer by going onto the driveway, as this to us would be private property. In any case, we didn't want to get in the way of the firemen, so we climbed through a hedge and skirted around the edge of a muddy and waterlogged field to the rear of the garden. I started to fumble in the dark for my camera with only the glow of the fire to help. As I got it out of the bag my fingers were so cold it was difficult to get the lens cap off let alone focus it. The change in temperature from my car to the outside had brought instant condensation to the viewfinder and lens. Just as I got everything clean a fireman came into view and as he walked around the rear of the shed several paint cans exploded. As they did, he crouched down and was silhouetted by the flare. I brought my camera up and, without attempting to focus, pressed the shutter.

In all honesty there was little we could photograph as we were out in the middle of nowhere without any street lighting and to use a flash would have been worthless. We were both so cold that we decided to head for home. Our feet were soaked through and frozen, as were our cameras. The following day I developed the black and white film at college and was amazed to find that the one reaction shot I had taken had come out pretty well. It was a very dramatic action picture, and I hastily took it into the Henley Standard to see if they could use it. To my amazement and great delight on the following Friday I had my first front page photo and credit to boot.

This arrangement with the firemen worked well, we got some pocket money, and they got some good publicity. We were always welcomed on the fire ground or at least tolerated, until one cold autumn morning.

I was getting ready for college and the pager burst into life just as I grabbed my toast which was ejected from the over enthusiastic machine. I looked at the clock and it was half past seven. This was a bit close to the cut off mark with being late for college, so I phoned Graham to see if the job was worth facing the wrath of my lecturers. He had already gone to the station window and the pad with the address on it also had the words persons reported. That was it, I was going to be late for college. I got down to the station in record time and met Graham who had his new video camera with him. This camera was the size of a bazooka with a battery pack to match. He squeezed into my mum's Ford Fiesta and off we went.

The address was over the border in Berkshire but the Henley

crews were the nearest, so they attended along with the Berkshire pumps. This would make things a little awkward as we would not be known by the other crews and would have to tread carefully. When we pulled up there was a different atmosphere. Both the crews were running about rather than their usual fast jogging, and they all, to a man, had very serious and concerned looks on their faces. There were no Berkshire crews there yet, but the police were there and were cordoning off the road. We started walking up the pavement and met Bob setting into the hydrant, (screwing a large metal pipe into the water mains in the ground). As we got up to him, he said, "You'd better stay well back and be careful what you film boys."

"Why what's happening?" Graham asked.

"There's someone missing and it's not looking good" Bob whispered loudly as he turned the hydrant on and watched the hose fill up. He then turned and ran back toward the pump.

We crossed over the now closed off road and made our way down the overgrown wet embankment trying not to slip into the half water filled ditch and staying as inconspicuous as possible. As we got closer to the row of terraced houses, we could see smoke coming out from the side of the far end house. Behind the front garden hedge, I could see hose reels and main hoses snaking down the garden path, and I could see the officer in charge crouching at the front door peering under the thick dark brown smoke that was rolling out the top of the open doorway.

On the lawn someone was setting up a first aid post and another was filling in the breathing apparatus control board. Graham started filming without using his tripod so he could move quickly if we were challenged by anyone. I got some general shots of the front of the house but for some reason I felt uncomfortable taking them and didn't want to move to get a better shot. I put my camera back in its bag and just stood quietly studying the developing scene. As I watched, a fireman wearing breathing apparatus came backwards out of the front door. As he appeared the officer at the side started to beckon the first aider shouting "Casualty!" The fireman kept walking backwards out of the smoke, and I could then see he was carrying the head and shoulders of someone. Then a second fireman appeared out of the smoke holding the persons feet. They moved over to the lawn and my view was then obscured by the hedge.

In the distance there were more two tones sounding and around the corner came the Berkshire engines and an ambulance. I looked over to Graham who had also stopped filming and was holding his camera down by his waist; he had the same fixed stare I felt I had. He looked at me and without a word being said we both turned and made our way back to the car. I dropped Graham off at work and as it was near to the Henley Standards deadline for news, I dropped the undeveloped film into the office on my way to college. As it was, I arrived only 20 minutes late and blamed it on the traffic.

That evening when I got home Graham phoned. He had been talking to the crews when they got back to station and was told that the lady, we had seen being carried out of the house had died. He had sent his video to the local TV station, and it was going to be on the evening news.

I watched the news that evening with mixed emotions, I was very happy for Graham as this was the first footage he had ever had on TV. On the other hand, I felt a little uncomfortable at profiting from someone else's tragedy. That Friday I got my second front page and began to understand the full story of what we had witnessed that morning.

The lady had been at a first-floor window only minutes before the crews arrived. She was seen to go back inside the house and wasn't seen again until the breathing apparatus (BA) team brought her out. Why did she go back inside the smoke-filled house? Why didn't she stay at the window in the fresh air? If it was that bad, why didn't she jump? I don't think we found out the answers to any of these questions.

I was coming to the end of my two-year college course and I was introduced to Ron, through a model I had photographed. He had his own company and was producing a multi-screen show called "The Venice Experience". He took me to London to see "The London Experience" to explain what he was trying to do. It was a 6-screen slide show for the tourists explaining the history of the city and where to go and what to see. Ours was to be installed in a cinema in Venice but his company was based in Twyford, not far from Henley. I had been recommended to him as a photographer / producer, and he offered me the job. However, he needed me to start immediately but I had 6 weeks and the final show to go at college. I approached my tutors for advice, and I will be eternally grateful for their response and support. They said that they were there to get us a career in photography, and I had been successful, albeit a bit early, they considered that my course work deserved a credit, but as I was leaving early, they would still give me a pass to get my diploma.

The job took me abroad for long periods so my activities with Graham and filming took a back seat for a while. Just after we got the show up and running, I was hit with redundancy and found myself back in the UK over Christmas 1985 working in the record department of WH Smiths. I wanted to keep some money coming in as I felt it was not right to depend on mum all the time.

In the January of 86, I received a phone call from the photography lab technician that I used to go to when working for Ron. He had heard of a company near Pinewood Studios who was looking for a photographer, so I gave them a ring, and after an interview later got the job.

They were a small audio-visual company who produced presentations for company conferences and such like. I was, again, the only photographer, and was very happy that on my first day I was taken to collect my new company car! There I was, still fresh out of college on about 10K a year and a company car. Things were looking good, and I was back in the world of photography.

With my fresh start I once again teamed up with Graham and started chasing the fire engines. The more I did this and saw the firemen at work, the more I began to think that "I could do that". I began to investigate what the entry requirements were to become a retained fireman. Retained, or On Call as it's referred to now, means that you must have a fulltime civilian job that will let you respond to a fire call when your pager goes off. They were nothing like what new recruits must go through nowadays. You had to have a basic education in English and Maths, Tick number one. You had to be able to carry a 12 stone man on your shoulders 100 yards in under 60 seconds, hmm, possibly tick number two, though not sure, and you had to have a minimum chest measurement of 91cm with a 5cm expansion, rather tight but possibly, tick number three.

I went down to the station on a Wednesday night, their drill night, to ask the Station Officer, John Gosby, if they had any vacancies? I had a long chat in his office about the job, but he knew I knew a lot about it already. I left with an application form to complete and return the following week. I completed the form as best I could, leaving blank the chest measurement, which was one of my concerns. I went back the following Wednesday. It was the 29th of January. I remember it well as the world was in shock because the previous day the Space Shuttle Challenger had blown up just after take-off. I was again shown into the office and met Divisional Officer, or otherwise known as DO Collins. We had a talk about the job and what it involved. Station Officer Gosby told the DO what I had been doing with my photography over the past couple of years.

We then went to the appliance bay where the same smell that I had first encountered on my Primary School visit all those years ago, hit me again. I remembered the oil, diesel, polish and above all the smell of smoke from the fire kit as we brushed past them. They were still hanging on their pegs as if they were on parade. Some of the crew were finishing their checks, whilst the rest were practicing the fireman's lift with Leading Fireman Ivor Gosby, the Station Officers brother, acting as the casualty. I remember seeing Ivor slumped on the floor by the wall and one of the crew trying to pick him up and put him over his shoulder. He could get him to waist height by jamming his knee into Ivor's crotch and then sliding him up the wall, but after that he was stuck.

I heard one of the crew say, "He's the most uncooperative casualty out. It's just like trying to pick up a sleeping cat". I remember thinking I had no chance at lifting him if even he was struggling. As we approached some of them greeted me, and Ivor stood up. I was told that he would be placed on my shoulders, and I would have to run around the drill yard and back to where we were in under 60 seconds. This sounds reasonably easy until I explain that the station was built on a very steep hill, so the first half of the run was downhill, but the second half was up hill. I made a conscious effort not to show any emotion but to just nod in understanding and ready myself for what was to come. Ivor stood on my right side and placed his right arm over my left shoulder. I was told to bend slightly forwards and place my right arm between his legs. As I did so he was lifted onto my shoulders, and I grasped his right hand with mine. I jiggled him a bit to get him more central on my shoulders and set off down the hill. I thought to myself he's not that heavy this should be ok. I decided not to run as I was frightened of losing my balance and ending up in a heap on the yard, I would walk fast instead. I ventured out into the darkness of the winter's evening, and I reached the bottom end of the yard quite easily and not puffing that much. I was quite proud of myself and turned around for the return journey. After a few steps it hit me.

The weight on me seemed to treble in an instance as I began the climb back to the starting point. My knees started to tremble a little and I was breathing more deeply than on the way down. My legs started to burn as I was determined to keep the same pace I had on the way up as I had on the way down. I felt Ivor begin to slide down my back as I began to really struggle, he whispered in my ear "come on Darran keep going, only 20 more yards" I finally made it to the big red doors and carefully put Ivor down the remaining few inches onto the ground. I stood bolt upright with the minimum of puffing and faced the officers, I was good at acting, I really wanted to collapse in a heap on the floor but that wouldn't look so good.

"Well done 48 seconds," said the DO. He let me catch my breath and then we went back into the office. He pulled out a tape measure from the desk drawer and he told me to put my arms up, this was the bit I was worried about. At the time I was a skinny 20-year-old, and the minimum chest measurement was going to be difficult to meet. I had dressed in a tee-shirt and shirt to pad me out a bit, you wouldn't believe it to look at me now. I was told to lift my arms up and to relax, the second bit easier said than done. Once he had taken the first measurement I was told to breathe in, this I did with gusto hoping to exceed the 5cm range. "Fine" was the reply as he whipped the tape measure back from my chest with the aplomb of a Savile Row Taylor, and I put my arms down. I looked over his shoulder as he bent down at the desk and a little smile broke out as he wrote 91 and 97, I'd done it!

I produced my school exam certificates and that was the end of the interview. I was told I would receive some forms in the near future and to make an appointment with my own doctor for a medical.

A few days later there was a knock on our door late one evening and Trevor Beales, the Sub Officer was stood there. "I've got your medical forms here, take them to your doctors as soon as you can". I read the forms, and they talked about lots of different measurements that I didn't understand, and I was beginning to worry that my dream was about to falter.

I had an appointment with my doctor the following week and as I sat in the waiting room, I could feel my heart thumping and I was beginning to sweat a little. This wouldn't help me a bit I thought. The light beside the doctor's name lit up and a buzzer sounded, here we go.

Once in the consulting room I handed the paperwork over to him. "Ah so you want to become a fireman then Darran? Ok then let's work through this lot then."

Half an hour later, after various prodding, coughing, deep breathing, reading of a sight chart and weeing into a paper cup, which was difficult as I was so nervous before I went in, I went to the toilet first and so struggled to get any out. He finally sat down and completed the remainder of the form. "Well, I see no reason why you can't be a fireman Darran, you've passed everything." I could have jumped up and kissed him there and then, but I thought better of it and just said "Thank you very much" with a Cheshire cat like grin.

After what seemed like an eternity a letter landed on the doormat with a franking of Oxfordshire Fire Service in red ink along the top. My hands were shaking as I carefully opened the envelope and pulled out the 8 or so pages. The first page with its black ink crest on the top had written underneath

Appointment as a Retained Fireman.......Darran Gough number 2675 to report to station B6 Henley on 7th May 1986

19:00 hours. It then went on to say what my annual retaining fee would be, my hours of cover contractual obligations and so on, but my eyes kept wandering back to "APPOINTMENT AS A RETAINED FIREMAN" I had done it!

I ran to find my mum and show her, much the same excited way that I had with my exam results all those years ago. She said she was very pleased for me, but there was a hesitation in her tone which I did not really pick up on at the time but I was to realise what it meant the first time I didn't come home when I was supposed to. I then phoned Graham to tell him the news. He was happy for me and jealous as, because of his build and eyesight, he could never apply. It also meant the end of our fire engine chasing partnership, but not of him taking photographs I would now become another grateful scrapbook recipient.

CHAPTER 3
"The Sprog Reports for Duty"

That Wednesday 7th May 1986 could not come around quickly enough for me. My contract had started on the 1st but that was a Thursday and I had to wait a whole week until my first official night at the station. I had arranged to finish work early and drove home along the M40 in record time. After a quick shower and tea, my attention turned to what I should wear. I decided on a shirt with no tie and smart trousers rather than jeans. I drove to the station and found a parking space at the bottom of the yard. As I walked back up the hill, I could see through the blinds that some of the crew were already in, so I didn't bother knocking, I just quietly opened the front door and went down the three stairs to the muster bay. This was a small lobby type room with, at one end, some heavy semi clear plastic strips hanging down across a double door size opening. The opening led into the appliance bay where the two fire engines were kept. In the middle of the floor was a low solid plinth, which was wooden around the sides and had a rubber top. Around the sides of the room were large black coat hooks with all the crews fire kit hanging on them. Over by the window were six separate pegs, three on either side of the window. On these pegs hung the officers' kit.

The kit of John Gosby, the Station Officer, included his white helmet which had one thin black bar running around it. Next to his kit was that of Mick Goddard the Sub Officer, Sub O or just Sub, second in charge, Trevor Beales having retired just before I joined. Clive Boulter and Ivor Gosby and Rick Williams the 3

Leading Firemen had their kit on the opposite side of the window. Gone were the black helmets I had seen as a child on the school visit. Mick had a yellow helmet with two black lines around it, and the Leading Firemen or LFFs had 1 black line. The thing that struck me about all their helmets except for the white one was that were all quite heavily battle scarred. Some had been obviously repainted, and Mick's was of an older style than the rest, as if he didn't want to part with an old friend. The tunics were also different. The longer serving had a black vinyl type jacket called a Teled. Inside this was a removable liner which was a poor attempt at a fleece. It was nicknamed a Hairy Mary because, if you had on only a tee shirt, you would itch like mad. In the collar was a label for the washing and care instructions and it actually bore the inscription "Keep away from naked flame"!

Below the tunics, on foot bars above the heating pipes, were the now yellow plastic leggings and the black rubber boots, all of which were clean and polished.

On the plinth in the middle of the muster bay was a large cardboard box which, bearing the word Gough 2675 B6 Henley, obviously belonged to me. I was dying to open it, but I waited to be invited to do so. People continue to arrive, and I was introduced to Graham Case, a new crew member. Graham had joined four weeks before me and was still finding his feet. As it turned out, we were to go through all our training together. This was a huge benefit to both of us, and we were to make a strong friendship.

18:30 arrived and we all went to line up in the appliance bay for the first parade. I stood at the end of the line and being still in my 'civvies', felt a little out of place. The Station Officer came out of his office and the Sub Officer called us to attention. This order was in no way military drill sharp, it was merely a smart move by all in respect to our leader. John told us all to stand at ease and then welcomed me to the crew with the words "he needs no introduction as you all know Darran, and he has swapped sides and joined us now". He then went on to detail who was riding which fire engine that evening and what was planned for the training session. I was to be taken by the Sub, who showed me around the station and kitted me out. We were called to attention and told "to your duties - fall out". We turned to the right, paused and broke the line.

The Sub took me back into the muster bay and then showed me the other rooms which led off from it. One of them was his office which, a little strangely, contained a shower. The other room had a sink, a scrubbing table, a store cupboard and some lockers. I was pointed to a locker which contained a key in its lock. This locker was to be mine. I was then shown "the most important piece of equipment in the building for sprogs", as Graham and I were known. That was, of course, "the kettle". No matter what the time of day or night, on return to the station following a 'shout', it was our job to get on a brew.

We went back into the muster bay, and, for me, it was rather like Christmas, as I was told to open the box. The excitement at seeing the gleaming brand-new fire kit was something I will never forget. The kit was all neatly folded and packed in plastic. The large helmet, a surprisingly lightweight, bright yellow cork item, in the middle at the front, bore the Oxfordshire Fire Service badge.

The sense of pride was overwhelming. I also experienced the same feeling twenty-three years later, when I became a full-time fire-fighter in the West Midlands, and with the rest of my course opened more big cardboard boxes. There, placed on top, was a rather more modern fibreglass helmet. It was somewhat heavier than the good old cork ones, but it was still yellow, and it still carried a badge. My feelings were nevertheless the same as they had been on that memorable occasion in Henley.

I was told to write my name and number on the inside and to find a peg on which to hang it. Next to Graham's kit, there was a gap in the long line of fire kit. Next out, was a pair of rubber boots. They had that lovely new rubber smell and were slightly sticky. They had loops on either side at the top to help you to pull them on, as well as oversized steel toe caps. Before I tried them on, I unpacked my new, bright yellow leggings and pulled them up over my trousers. They had one long elastic brace that I slipped over my shoulder. I then had to feed my feet into the boots with my trousers going inside and the leggings on the outside. Mick then looked at the length of the leggings compared with the boots. I adjusted the brace so that I had no more than one inch from the bottom of the leg to the ankle of the boot.

I then pulled one of the two tunics out of the box. This was not the teled type; it was called a Mountain Tunic. It was made of a heavy black serge material with two rows of three large silver crested buttons and a long wide strip of Velcro. I pulled it on and did up the buttons and Velcro. The collar was a bit rough and had a flap which came around the neck. I was told to bend over, to crouch down and to see if there were any tight spots. Having found no tight spots, I took off the tunic and hung it on the peg under my helmet, with the buttons facing towards the wall. I then slid my leggings down over my boots and stepped out of them. Using the handles, I placed them on the warm pipes under my kit and was told to put the brace inside and to fold the top over the boots like the others. Returning to the box, there was a neatly coiled up piece of rope, which I was told was my belt line. I was to keep this in the inside pocket of my tunic, as we were no longer issued the belts from which they used to hang them. Next, was a large square piece of black silk type material which was to be worn around the neck inside the collar.

I was then given a piece of simple kit, which is still issued today. In some brigades it is a disciplinary offence if you lose this item, THE ACME THUNDERER WHISTLE! Made of stainless steel, and used by football referees, it was to be fastened to the inside pocket of my tunic. It was to be blown repeatedly in three short bursts, to signal a possible building collapse and to evacuate the building. This was the countrywide universal evacuation signal and could be made by anyone on the fireground.

The final piece of firefighting kit was a pair of light blue rubberised gloves. They were called Plastichrome and they resembled a pair of ordinary gardening gloves. Once all my fire kit, known as Personal Protection Equipment (PPE), was hung up correctly, I returned to the box which was still half full. Inside was my uniform, all neatly packed and folded. I was told to go into the locker room and to try it on. This came with the Sub's warning, "I'll be very surprised if it all fits first time". I heaved the large box through the small doorway, and the Sub went back to his office. In the box were.

Two light blue long sleeve shirts

Two pairs of black heavy serge trousers with a funny thin clip style belt.

A dark blue heavy woollen NATO style jumper.

A black tie.

Three pairs of black 100% nylon socks.

One pair of black slip-on shoes.

One peaked cap

One cap badge

One undress jacket

I tried on one of everything and came out of the locker room feeling like a tailor's dummy. The shirt was heavy with starch and was creased where it was folded. The jumper felt two sizes too small, but I was assured it would give and stretch after a couple of washes. The trousers were a good fit but needed some work on the crease. The socks were awful. They were 100% nylon, which seemed a bit odd for the fire service, and I soon bought replacements of my own. The peaked cap had a thin red pipe around it, and I was shown how to fix the silver and enamel badge on it. This same badge was mounted on a silver, ash handled, fireman's axe that was presented to me by my crew when I retired from the retained in 2013. The shoes were a good fit and, I thought, would be quite comfortable when worn in. I left the duplicate items in the box to take home and, yes, ask mum to wash for me!

The Sub then sat me down in his office and began to tell me what to expect over the next eight weeks. There was a set list of topics to cover, and the amount of time allotted to them. It was explained that, once I had completed the list, I would be tested by the Divisional Commander, D.O. Collins, who, if he was happy with what he saw, would put me "On The Run" I was also given a piece of paper listing the topics. I was to get this signed up by whoever took me for the training. The whole purpose of this was for me to become a 'gofer'. I stared blankly at Mick, not understanding what he meant. He then said, "Go for this, go for that," ah, I had got it! Once on the run I would, in the next 6 months, spend a week at the training centre at Didcot for my Basic Training Course.

The list included 12 hours for:-.

Knots and lines

Reef Knot	Rolling Hitch
Sheet Bends	Round Turn and Two Half Hitches
Clove Hitch	Timber Hitch
Catspaw	Bowline
Chair Knot	

I had heard of some from my time in the scouts, but others sounded "double Dutch". I would also have to learn the location of every piece of equipment on the fire engine and what needs to go with what. I needed to learn the hand signals, how to put up all the different types and sizes of ladder, as well as the safety procedures to follow when attending what was then called a Road Traffic Accident (RTA).

This training would take a minimum of twelve hours i.e., six drill nights. The other two drill nights would be taken up with learning about how to book on and off duty as well as how to recognise the senior ranks, either in their uniform or in their fire kit. A tip which I was given was basically to address as "SIR" anyone who wore a white helmet. I had to understand how the Brigade's orders were published and located. Then there was the more practical side. I was told how a fire call was received. Even though I was not yet on the run, I was given an alerter, or pager unit, to take home and to get used to carrying around. When operational or 'on call', I had to always have it with me. Every evening at 18:45 it would go off with a test call. This was a series of long bleeps. I remember them well as they always used to go off during the commercial break while my mum was watching Crossroads.

A fire signal was different. It was a short bleep followed by a long bleep then by a short bleep and so on for twenty seconds. Then there was a pause, and it would repeat the sequence. If we didn't get the test signal, then we were to phone the Station Officer. I was then shown where to park my car when responding to a fire call and it was impressed upon me that, always when responding, I MUST obey all the rules of the road, with no exceptions. As the person, whom I will not name, was telling me this, he showed me his crossed fingers and I got the message.

I was then told about the tally system. On the appliance bay wall was a board with twelve hooks in it. On the hooks were twelve square tallies each being about one inch by two. They were in two columns of six, one for the Ladder, which is for the first fire engine out of the station, and the other for the Pump, or second engine. The top tallies in each row were white and said on them either Ladder or Pump, with OIC (Officer in Charge) under that. The next was red and said Driver. The next two were yellow and said BA, short for breathing apparatus. The last two were green and simply said five and six. When I came into the station on a shout, I was to take the highest tally up the column I was entitled to. This then usually ensured I was riding that fire engine for that call. Once on the run, I could only take one of the green tallies. However, depending on how many men were on call at that time, you could be told to stand down from the engine to ensure that there were enough men left to crew the second appliance, should that be required. As I progressed through my training, qualifying in wearing the breathing apparatus (B.A.), passing my HGV driving course, I could then take the tallies which were higher up the board.

Mick told me that, for training purposes, I would be paired with Graham so I would have four weeks in which to catch up with him. I was given a station key and a small red official fire service drill book for me to study I was told that I could come down to the station at any time I wanted to, so I could look around the appliances and to learn the location of everything. Mick said that he would be more than happy to come down as well, at weekends, to help me and Graham with extra training so that we could get on the run quicker. The hidden agenda for this was that, as soon as we were on the run, it meant that the rest of the crew could take some time off, if crewing levels permitted it.

Mick looked at the clock above the printer. It was already twenty past eight, and the drill evening finished in just ten minutes time. I wondered where two hours had gone? Everyone was coming back into the muster bay and was hanging their kit back up on the pegs after their training session. I couldn't help thinking that all their kit, although clean, looked used, whilst mine, virgin and pristine in the middle of the row, stood out like a sore thumb. I couldn't wait to get it just a little bit dirty so that it didn't stand out so much.

As the clock reached half past, individuals started to make their way up the stairs to the social club on the top floor. At this time, all fire stations had a social club, where the retired members and the families could meet up for an evening. At Henley, the Wednesday night tended to be the domain of the current crew, and they all took it in turn to run the bar. The club was also open on Friday and Saturday nights and on Sunday lunchtimes. At those times, Tony's sister Pam, ran the bar in return for a small fee. Noticing that the officers were among the last to come upstairs, I supposed that they were completing paperwork or some such like.

This was the first time on that evening that I had been able to mix with my fellow crew mates. To be a retained fireman in the 1980's, you had to be in full time employment. That is why I hadn't been able to join whilst I was at college, even though I had tried a couple of times.

Station Officer John Gosby had been a full-time fireman in Berkshire but now had his own window cleaning round and had been in the service for nearly 20 years.

Leading Fireman Clive Boulter was a self-employed builder.

Leading Fireman Ivor Gosby, John's brother, worked for the Council.

Sub Officer Mick Goddard drove a rubbish lorry for a private firm.

Leading Fireman Vic Warner worked for Stuart Turner's, a Henley firm of pump manufacturers.

Then, we had the firemen. Bullet Trendall worked with Mick on the dust carts. Tony Buckett was a mechanic. Bob Dobson was a lecturer at the local college. As for Nick Evans, I don't know what he really did at that time, although it was likely to be "a bit of this and a bit of that". Simon Cook worked at the Management College, on ground maintenance. Graham Case worked for Brakspear's Brewery in the town. Most of them had good solid practical skills and, with this rather eclectic mix of trades, you could usually depend on someone knowing what to do in an emergency.

I sat on one of the low comfy seats and Graham came over. He told me what he had done so far in his first four weeks and we, pardon the pun, got on like a house on fire! Before I left to go home, we had arranged to meet up that weekend to practise our knots and lines and to look over the pumps and to try to learn where all the kit was.

After about an hour, I made my way to the door and said my goodbyes. I collected my half empty cardboard box, containing my alerter, drill book, and civilian clothes and I made my way down the drill yard to my car. I had a strange feeling of finally being somewhere where I belonged. Don't get me wrong, at the time I really enjoyed my photography work, but there was something missing. I wasn't sure whether it was a sense of belonging, possibly to a team. However, I was sure that I was mentally knackered, but I was very happy.

I got home and dropped the box onto the living room floor. Mum looked both at me in my uniform and at the still folded duplicate items in the box. "I suppose you want that lot washed?"

she said in a slightly sarcastic tone. "Thanks mum" was all I could really say. I said my good nights and retired to my room. Carefully, I hung my cap on the hook on the back of my bedroom door. After hanging up my civvies and uniform, I propped myself up in bed and read my new drill book, looking at the numerous drills laid out in pictures which detailed what each numbered fireman did in each part of the drill. It looked extremely confusing and, as I flicked through the pages, my eyes grew heavy, and I drifted off into a deep sleep.

CHAPTER 4
On The Run

The next day at work was a quiet one. We were in-between jobs and waiting for the material to come in to make a presentation. I decided to make some stock photos for our catalogue. Whilst in the darkroom, with the no entry light on outside, I got somewhat distracted with my new Fire Service drill book. I started flicking through the pages.

Two hours later, whilst waiting for the film to develop, I found myself with a wet film and no prints. I hadn't made any plans for that evening, and after tea, I found myself outside the front door of the station, my station, letting myself in.

With nobody about, it seemed an eerie place, silent apart from the whirring coming from the cooling fans in the printer cabinet. I felt a little bit naughty, as if I was somewhere I shouldn't be.

I had every right to be there, and I had been told to come down as often as I liked to learn the equipment. So that's what I did.

The fire engines they had at Henley were Bedford TK models. To the public they would look the same, but they weren't.

One had on it, both the larger 13.5m ladder and the Road Traffic Accident (then called RTA) equipment. This engine was usually the first out of the station. It was called Bravo Six Lima and had B6L in the windscreen. The B stood for B Division, the 6 was for Henley and the L was for water tender Ladder.

The second appliance had a smaller 10.5m ladder on it and

did not have any cutting equipment. Unless the call out was for a minor fire such as grass, rubbish or maybe a chimney fire, this appliance was usually the second out of the station, This was called Bravo Six Whisky, or B6W. The W stood for Water tender, even though both appliances had the same amount of water and size of pumps on board.

I climbed into the rear cab of Lima and sat on the bench gazing at all the equipment. It had two breathing apparatus sets in the middle, which were secured in place by a leather strap with a belt type buckle that fell apart when you flipped the top up. Above them, on shelves, were the reflective jackets and two metal cone type horns, much like those you get at children's parties.

but these were for use on the railways by safety officers. There were other small bits of kit that I would get to know later. I daydreamed for a while, staring between the two vertical chrome poles between the rear and front cabs, imagining what it must feel like to be on a call, or shout, and how soon it would be until I found out

I climbed down backwards, as I had already been told and went to look in the lockers. The locker shutters on the engines were still all red, not silver as nowadays, and were channelled, a bit like corrugated iron.

I slid up the first one, which moved effortlessly and was met by the Lightweight Portable Pump, or LPP. This was a small car engine mounted in a tubular frame with four folding handles, one on each corner. The name lightweight was a bit of a joke as it took four men to lift it and, even then, you couldn't carry it more than 50 yards or so before stopping for a breather.

It was started with a pull cord, much the same way as a lawn mower. However, it was not like a lawn mower, as if you didn't get it primed and started in the first two or three goes you started to run out of puff as you were pulling a four-stroke engine, not a little two stroke. It had two outlets with large knurled wheeled valve handles that was where the hoses were plugged in. At the same end, but at the bottom, it had a large opening covered by a large chrome cap that, when it was taken off, connected to four lengths of hard suction, which were kept on the roof of the engine.

This hose was about 3 yards long and rigid, with two large screw fittings, one at either end, one male and one female. They were very heavy, and to connect them together you held one coupling between your legs and someone held the next length of hose in the same manner. You then shuffled together and once the threads were aligned, you used a pair of suction wrenches to screw the couplings together. There were four of these wrenches on the engine, two on the LPP and the other two in the pump bay, which was the locker at the back of the appliance. They were there because this hard suction could also be connected to the main engine pump. On the final length of the suction was a metal strainer. This was full of holes to stop stones etc from going up the pipe and damaging the pump.

During my training I was given lots of information that I would not describe as being useless, but certainly as being not very useful. However, I have never forgotten one bit of this information. It is that if you add the total diameter of all the little holes then it equals to the diameter of the hard suction. Therefore, there is no reduction in the amount of water being taken from a river.

Once all the lengths of suction were coupled together you had to put a wicker and canvas basket over the metal strainer. This basket has been in use for many, many years, and is still in use today in various forms. It is true what they say that "if it isn't broken, DON'T FIX IT". The basket was to stop the metal strainer from sinking into the mud and clogging up. Once this was on you then had to use a line, not a rope, to secure the strainer to the suction, using a clove hitch. This line was then used all along the hard suction, with a few half hitches, to take the weight off the couplings, and fastened to the pump.

The other end of the line was then used to lower the strainer into the water and to position it so that, if there was a flow in the water, it was facing upstream to prevent any vortex's forming and entraining air into the pump.

All this long explanation was how to get the LPP to work, but I had seen them do it for real at the boat fire and it looked effortless. I would have to get to that standard, and I would be tested for real sooner than I expected.

I continued to browse through the lockers at the various tools.

Amongst them was a sledge- hammer, or the universal door key as Tony called it! There were bolt croppers and saws of all different shapes and sizes.

In the middle locker, above the wheels, was the hose. Gone were the canvas hoses that I had seen on my school visit. They now had rubberised bright red hose of two sizes, 1 ¾ and 2 ¾ inch diameter. The reason for two different sizes was that the 1 ¾ was easier to manoeuvre in tight spaces, i.e. in a house.

Also, if you had limited water supply, as in most countryside areas, then it used up less water than the larger 2 ¾. The larger hose was rolled up with the female coupling in the middle, with its two hand lugs sticking out each side to help carry and run out. The male coupling was on the outside on top. There were fourteen lengths of this hose, seven on each side of the engine. The smaller hose, of which there were four lengths, two on either side, was rolled up in a different way.

It was called 'Dutch rolled'. During the makeup, putting all the equipment back on the appliance, the hose was under run. This involved uncoupling the hose and putting one of the couplings over your right shoulder and walking forward, feeding the hose over your shoulder as you go. This emptied out any residual water in the hose, after which it was folded in half, lengthways, with the female coupling on top, approx. a foot and a half behind the male.

Then one person starts to roll the hose from the middle towards the couplings whilst a second holds the top half of the hose, facing the first but walking backwards, keeping it tight as it is rolled. This is a bit laborious, but the finished hose has both couplings together on the outside. The idea is to hold this hose with the couplings at the bottom in your right hand with your left steadying the top of the roll. You then bowl the hose out, keeping hold of the couplings, and, when it's near the end of the roll, you whip your hand back quickly to finish the job.

In the pump bay, at the rear, was another method of storing

hose, the strike jet. This hose was flaked from side to side on a shelf on top of the pump casing. Already attached, one end had a branch, which Americans would know as a nozzle. When you arrived at a job and you needed a lot of water quickly, you would grab the strike jet and run with it.

It had two lengths of hose attached and would simply flake out as you ran. When the end neared, the pump operator would grab the end coupling and snap it into the pump. Finally, in each rear side locker was a hose reel on a drum that is permanently connected to the major pump. This is a long black hard hose, a bit like a garden hose, but can take high pressure and is the chosen first attack hose. There were 60m on each side in 3 lengths of 20m. Either side could be detached to extend the other to 120m. They had a permanently attached hose reel branch, or gun, which would either deliver a narrow jet or, at the twist of a handle, turn it into a spray.

Without setting into an underground hydrant, or open water supply, the 400 gallons carried by each appliance could last around 20-30 minutes if a hose reel was being used, but, if a main jet 2 ¾ hose was in use, and the pump operator wasn't paying attention, the tank could empty in just less than 2 minutes.

I looked at my watch and I had, again, been in the station for nearly two hours, so I shut all the lockers I had been in, turned the lights off, and made my way home.

Over the rest of the week, I went down to the station every night for an hour or so. Sometimes, one of the JO's, or junior officers, would be there completing some paperwork or such like. I liked this because they could see I was being proactive in gaining knowledge, and because I could ask them questions about pieces of kit. Sometimes, I could see I was being a little annoying and was distracting them from their work, so I just amused myself, so to speak. I met up with Graham on the Saturday morning and we spent the time practising our knots and lines. He showed me which knots were used to tie on certain pieces of kit. Even though he wasn't really supposed to, he started to show me drills he had already been taught in his first four weeks, so I could catch up, and we could both do our 'Competency to ride' test together.

We went through how to haul a sledgehammer and large axe aloft using clove hitches and half hitches, neither of which, in 28 years, I have had to use. The one knot we practised repeatedly was the rolling hitch to the left or right. I have used it many a time. This knot is used to haul hose aloft and to secure it when working at height.

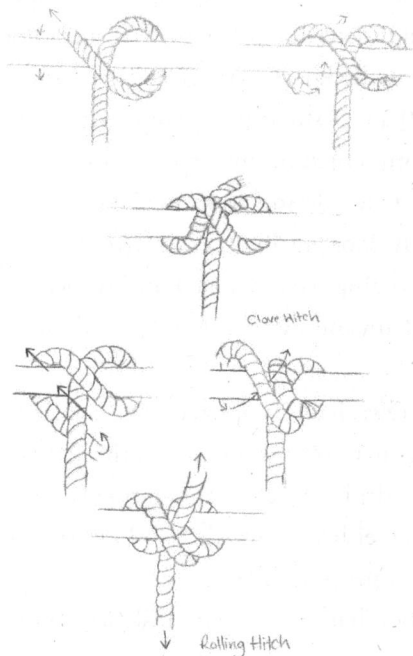

Clove Hitch

Rolling Hitch

If you don't start it correctly, throwing the line across the hose in the direction the water will flow, then the knot comes out backwards and will not support the weight of the hose when it is filled with water. There wasn't an awful lot we could do on our own, as most training needed at least a crew of four with one officer. We did, however, play hide and go fetch. I would name a piece of equipment and Graham would have to go and find it on the appliance, and then say all he knew about it and how it worked. Then it would be my turn to go fetch, and so on. This game worked well, and, by the next drill night, I was almost up to speed with Graham.

That following Wednesday I had to concentrate on not constantly smiling like a Cheshire cat as I walked into the station in uniform for the first time. I had spent a couple of hours on my shoes bulling them up, as I was a Warrant Officer in the Air Cadets and knew the knack for a glass like finish.

It was 18:30 and I stood in line on first parade. At the end, John Gosby commented on my shoes and the fact that most of the crew could take a lesson from me. "Great second night", I thought, already looking like goody two shoes, literally. The crews for the evening were detailed and we were dismissed for our duties. Graham and I were told to get our kit on and to meet the Sub outside.

Once the crews had completed their weekly checks, they joined us. We were to start drilling with the 13.5m ladder. Now, in all fire service drills, the crew are numbered one to either five or six, depending either on the drill to be carried out or on the actual number of people available.

Each number had their own task and position on the pump.

1: OIC – Officer in charge

2: Driver – Pump Operator

3: Sat behind the OIC

4: Sat behind the driver

5 and 6: Sat in the middle.

On a 13.5 drill you needed a minimum of four people. Where you were sat on the appliance corresponded to where you were positioned on the ladder, with No.1 giving the orders at the top, or head, of the ladder, No.2 stood opposite him at the heel, No 3 opposite No.1 and No. 4 opposite No. 2 on the heel, or Jack Beam which is the bottom of the ladder.

The Sub said we would very slowly walk through pitching the ladder to the third floor of the drill tower.

We were detailed our positions: Ivor No.1, Graham No.2, Tony No.3 with me No.4.

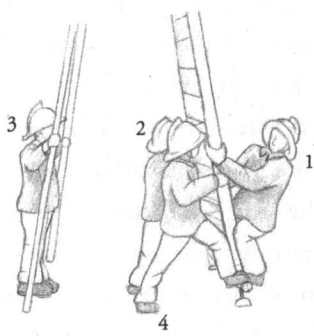

On the command "Get to work", we dismounted the engine and walked to the rear. Within a couple of weeks, we would be running.

There now followed a sequence of choreographed moves that any ballroom dancer would be proud of. We all moved into different positions and carried out actions to a set pattern to get the ladder off the fire engine and extended up the drill tower. After which we then did all the moves in reverse to get the ladder back down and onto the fire engine.

We repeated this drill a few times with us all taking turns in different positions on the ladder. So that the old hands wouldn't get too bored, the Sub then took just myself and Graham around the appliance for "some Q and A". He was very pleased with our knowledge. Our games of hide and go fetch had paid off.

The next few weeks followed roughly the same pattern. I would spend most evenings at the station, sometimes meeting up with Graham and, on the weekends, the Sub as well. Once when we were on station, a shout came in. The sound of the bells going down made us jump and we had to put the equipment quickly back on the pumps. We did this just as the first man came through the door. We had already had a sneaky peek at the printout and saw they had an RTA near Stonor. It was very frustrating just having to stand there watching the crew run in, grab their tally and fire kit. The first four or five in had time to get dressed properly in the muster bay, although the driver, was not allowed to drive in fire kit, so he had to stuff it into the LPP locker or throw it in the rear cab if there was space. The later ones had just had to grab their kit and get dressed en route. I would become a victim of this on my second shout. More of that, later.

The engines roared into life and we were left in an exhaust filled empty appliance bay, looking at each other wondering what to do next. After a pause, we both came up with the same idea at almost the same time. We jumped into my car and followed them.

As we drew near to the incident, a policeman flagged us down. We were wearing our fire service t shirts and explained that we were trainees and that we were at the station when the call had come in. We asked if we could go through and watch what was going on? He let us through on the promise that we would not get in the way.

I parked the car out of the way and walked slowly up to the scene. It wasn't that busy. There were no people trapped and the crews were just mopping up a fuel leak. John Gosby saw us and beckoned us over. He asked us what we were doing there. He was happy when we explained that the police had let us through. He showed us what they were doing and explained the procedures which they were carrying out.

On the following drill night, we were told that we had completed the work up program, and that everything was signed off. The Divisional Commander, DO Carver, would be down on the following Wednesday to carry out our Competency to Ride Test.

The rest of the week dragged but Wednesday came around eventually. I left work early and was home in time for an early tea and to grab my uniform that mum had specially washed and ironed. My shoes had been given the treatment the night before and I was ready to go.

I got down to the station a bit early and spent the time

cleaning my fire kit, which was really a waste of time as I had already cleaned it that weekend.

People started to arrive and were generally milling about. Graham came in and we had a chat. We were both giving off a bit of bravado saying we were OK and ready, but I could tell that he was just as nervous as I was. I saw the 'DO come in', and he went straight into the station office and closed the door.

1830 came and we were all stood in a line to the rear of the two engines. John came out with DO Carver and then the Sub called us to attention. We were introduced and were told what was going to happen that evening. To everyone else this was old hat as they had been through it numerous times before, but still they didn't want to be made to look stupid, or to slip up.

The pre drill checks on the engines had already been completed before parade, rather than just after, so we were ready to get started.

Graham and I had to put on the reflective RTA jackets so that the DO could pick us out easily from the rest of the crew.

The DO took us both to one side and explained what he was looking for and the standards he was expecting. He wasn't looking for perfection as, with only 16 hours of official training behind us, that would not be realistic. Neither was he looking for speed. What he was looking for was a safe and confident session, and thus we would not be tested on anything that we had not covered already. Having heard his calming talk, which had not calmed us, we were off.

While we were having our talk, the pumps were pulled out onto the drill yard, the lads were standing behind them, waiting for us. We were told to join them, one in each crew.

The first drill involved Graham's crew. They had to pitch the

13.5m ladder to the drill tower. While they were doing this, my crew was told to stand to stand to one side. I stood watching him and wished that I was the one to start first. The DO was stood with John, watching from just inside the appliance bay. He wasn't taking notes, he just had the sheet of paper that detailed all the training we had done and would be slowly working his way through the list with both of us.

Graham's first drill went smoothly, and the ladder was re-stowed on 6 Lima. My crew was called forward. "Here we go," I thought.

John detailed our drill. We were to slip and pitch the 13.5 ladder to the third floor of the tower to the right-hand side of the window. E "Fireman Gough you are number 3 in the crew", he said. Number 3, I thought, that means being on the pole taking the ladder off the back of the appliance on the nearside. "OK I'm happy with that ", I concluded.

"Any questions?" We all remained silent. "RIGHT, AS DETAILED, GET TO WORK" he shouted.

We jumped to attention and turned to face the back of the ladder. Ivor and Tony reached up to the ladder releases and Ivor called "Prepare to slip, SLIP". With this, Bullet and I reached up and caught the jack beam of the ladder as it slid down off of the gantry. When we caught it, we turned toward each other and carefully swapped hands over to carry the ladder to the foot of the tower. I sighted the foot 1/3 the working height from the wall and up it went.   I was beginning to sweat a little and my gloves slipped a bit as I pulled on the poles. Grip harder I thought, but the ladder went up relatively smoothly. It was extended and put into the window in almost the right spot. It was a bit to the left but was well within the realm of acceptability.

We all stood smartly to attention and, following a nod from the DO, John called "KNOCK OFF, MAKE UP". With this, we brought the ladder back down and put it back on the appliance.

Next, was a combination drill. This involved both crews and Graham and I were detailed specific jobs in the drill and the others were told not to do them.

The "get to work" was called and we were off.

Graham's crew pitched the 10.5m ladder to the tower and with Steve Clarke, he went aloft with a line. My crew had to get the pump to work and supply a hose line to the crew aloft. While Mick went and set into the hydrant, I had to run out two lengths of hose and flake it at the bottom of the tower, in readiness to tie the knots so that Graham could haul it aloft. Having just run out the second length of hose and snapped the branch, I heard Graham shout from the tower, "STAND FROM UNDER" Glancing round quickly and seeing no one near, I shouted back "CLEAR". With that, there was a dull thud as the line bag landed on the concrete a few feet away from me, with the line trailing back up to the window where Graham had tied it off first. I grabbed the bag and paused for a second. "Now throw the line in the direction the water will flow", I muttered under my breath. To do this easily, I was stood the wrong side of the hose, so I hopped over it quickly. I felt the eyes of the two officers burning into the back of my neck as I began to tie the rolling hitch.

Over once, over twice, half hitch. It looked good to me. Now, grab the line about 6 feet from the hitch and hold it up in front of me and make two loops, one towards me with one hand and, using the other hand, make another loop but this time away from me. Right, now bring the two loops together. There, that was a clove hitch. Now, slip it over the Nobel branch and pull it tight. I looked at my efforts and it looked all right. Mentally, I breathed a sigh of relief and shouted, "HAUL ALOFT". The line tightened and the hose jerked upwards, like a snake for its charmer in a Calcutta Street market. I stood, almost hypnotised, watching it go up the side of the tower and into the window. The knot had held, YES! I stood watching the window and Graham's helmet bobbing frantically up and down from the sill, as he made the hose secure. Finally, he looked out of the window and raised his right arm up vertically and shouted, "WATER ON, NUMBER ONE DELIVERY, FIVE BARS PRESSURE WORKING FROM THE THIRD FLOOR". I raised my right arm in acknowledgement and shouted back his order.

I then turned and began to follow the hose back to the pump. When I reached the final coupling of the hose before the tower, I bent down and connected it to the rest of the hose. This final coupling was always left uncoupled until "water on" was requested as a safety measure so that if the pump operator inadvertently turned the water on too early, the men on the ladder would not be blown off. I arrived at the pump and facing Tony, who was the pump operator, I raised my right arm again. Even though he was only a couple of feet away from me, I shouted at the top of my voice the order for "water on". He raised his arm and shouted it back to me. Now all this repetition may seem a bit long-winded to an outsider, but it helps confirm the orders are being received correctly. Yes, on the real fire-ground, the orders are not as snappily shouted. However, we did have a DO watching us, so a bit of exaggeration would not go amiss. Tony turned to the pump and slowly opened the screw down valve on the left most delivery, number one. The hose started to inflate, and I followed it back to the tower, kicking out any kinks that occurred. Once the water was streaming out of the branch in the tower, Tony increased the pumping pressure to the required 5 bars. I saw John and the DO talking and nodding together. I couldn't hear what they were saying as the roaring of the pumps drowned out any normal conversation. This is another reason why orders are shouted.

John turned to us raising his arms horizontally and then brought them smartly down to his sides. As he did so, he again shouted "KNOCK OFF, MAKE UP"

With this, we all burst back into action, shutting down the pumps, retrieving the hose from the tower, making up all the equipment and re-stowing the lines and ladder back on the appliances. This was all done at an efficient pace by all of the crew except for Graham and me. As we were conscious that we were being watched, we did everything at full chat and, with all the fire kit on, we were sweating like someone on a sun kissed beach. However, this was Henley on a typically cool July evening.

We did a couple of more combination drills with Graham and me frequently swapping roles so that we both covered all which the DO wanted to see. When he was happy, he sent the rest of the crew inside to cool down and to carry on with whatever else needed doing on that evening. Graham and I stayed outside with the two officers. The DO told us to relax our dress, and to go and get a quick drink. As we took our helmets and tunics off a rolling fog of steam and sweat enveloped us both. Our tee shirts were ringing wet and our trousers under the plastic leggings weren't much drier. We downed a quick glass of water and returned to the officers who were deep in conversation in the appliance bay.

The DO greeted us and said that he was very happy with what he had seen, so far. He was now going to ask us some questions about the equipment and its operation. There then followed 20 minutes of the 'Go fetch game' that we had been playing on each other the previous weeks. We both seemed to get through that without too much bother and were both able to explain the safety words of "command", "rest", "still", "stand from under " and "carry on", and who could call them.

At the end the DO said "I am very happy with both of your performances, and this reflects on the quality of training which you have had from the crew here". He gave an approving nod towards John, who smiled courteously in return. "I am to understand from your Station Officer that it is also due to the exemplary amount of extra training which you both have put in off duty. Well done." Graham and I looked at each other and shared a little grin. "I will instruct the Station Officer to issue you with your alerters". John, who was standing slightly behind the DO looked straight at us and, with a slightly concerned look on his face, shock his head quickly to indicate that we must say nothing. This was because he had already given us the alerters some weeks ago so that we could get used to carrying them around with us. Whilst this was not against any rules, it wasn't common practice, and we would avoid any awkward questions being asked of him if we stayed schtum. The DO continued, "I am pleased to say that you are both on the run, as of midnight tonight. Well done and I'm sure you will both be valuable assets to the station." With this, he shook our hands as did John whilst saying a quiet "well done". With that, the officers left us and went into the office to complete the paperwork whilst we went to get out of our leggings and to clean up.

As we entered the muster bay, it was empty of people. All of the fire kit was hanging up on the pegs and any equipment had been tided away. I looked at the clock. It was 2130, half an hour past our stand down time. We put our leggings and boots back on the rack, and went upstairs to the bar, which was now inhabited by the crew.

"Well?" was the almost synchronised question as we walked through the door. With grins as big as a Cheshire Cats', we both replied, "On the run". I expected a little cheer or a well done but, no, the first reply was, "drinks on you then". Looking back, I now understand the look of resignation on a couple of their faces. With we two going on the run, it meant that some of the regulars would miss out on more shouts as we lived nearer to the station than they did. This would reduce their wages. Certainly, 15 years later, when I was in charge of my own station, I used to have a nonchalant look at any new applicant's home or responding address to see if they were nearer to the station than I was. It never stopped me recruiting them if they were the right person for the job. As I was the officer in charge, their time to station and the order through the door would rarely, if ever, prevent me from riding the engine. However, I would be able to predict who from my crew would be bumped off and be ready for the resultant grumbles.

Back in Henley, this would not have an immediate effect on the crew, as Graham and I were the probies and as such were only entitled to take one of two green tallies per appliance.

So, they would still be able to bump us off the truck if specific skills were needed, i.e. a breathing apparatus wearer or a driver.

We both put some money behind the bar and settled down to the rest of the evening chatting about football, playing darts, girls etc. A while later, John and the DO came up and joined us for a drink. The DO had a quick one as he didn't want to be rude. He could not stay longer as he was on call and had to get back to Oxford sooner rather than later.

The room started to thin out and I decided to head home as well. When I got in, Mum was still up and had that expectant look on her face waiting for my news. When I told her that I had passed and was on the run, she was genuinely pleased for me, but there was the same hesitation in her manner which had been there when I received my letter of contract a couple of months ago.

I gave her a kiss and said goodnight. I went to my room and began what would become a normal bedtime routine for me over the next 28 years. As I undressed, I laid out my clothes along the floor in a pattern leading from my bedside to the door. Pants, tee shirt, trousers, socks and shoes. This would change later to pants, boiler suit (now known as a onesie), socks in top pocket and shoes (or boots, when I used a moped to respond). These changes were all made to speed up my time in getting to the station, but for now, I stayed awake until midnight. As the pips came over my bedside radio, I began both my operational career in the fire service and what turned out to be, or seemed to be at least, a very long wait for my first shout.

CHAPTER 5
Now the Learning Begins

"BEEP BEEP BEEP BEEP" - I woke up, with a start, to the sound of my radio alarm rather than that of my alerter. I then did what every retained fireman does on a regular basis - I pressed the button on top of the alerter, just to make sure it was working. It flashed its acknowledgement to say that all was well.

I got up, dressed and had breakfast. All of the usual routine, except that, when I went to my car, I subconsciously slipped into another action which would haunt me for the rest of my career in the retained. The usual man thing of a pat on the back trouser pocket to confirm his wallet was there, followed by a front pocket pat for keys continued with a pat on the waist in order to confirm that I had my alerter. Years later, this would be followed by a fourth pat for a mobile phone. Because I worked over 30 miles from home, I really didn't need to have my alerter with me, but this blue plastic mini-brick sized box of tricks would become my shadow now and I would have it with me 24/7. This was just in case, on my way home from work, when I got within range, about 5 miles of the transmitter on the station's drill tower, a call came in then I would at least have a chance of catching a pump, rather than missing it if my new friend was left at home!

When I got to work, my employers asked me, as soon as I got in the door, how things had gone on the previous night? I showed them the alerter. I had kept it hidden, as I had said eailier. John had given Graham and me the units a few weeks before in order for us to get used to having them around. I worked for an elderly husband and wife and they did seem genuinely interested in what I had gone through. I was brought back down to earth with the work schedule for the day, and off I went to the darkroom.

I must admit that, once again, my mind wasn't really on the job. I was wondering when my first shout would be. What would it be? Had Graham already had his first shout? Graham worked for the local brewery, Brakspears, right in the middle of town. The brewery has long since gone and severely overpriced penthouse flats now inhabit the spaces left by the vats. You can still buy Brakespears today, but it is brewed by a conglomerate company in Witney, and it doesn't taste the same. The old Henley residents still reminisce about the brewing days, Mondays and Thursdays, I think, and about the Marmite, the smell of the brewing. You either loved it or hated it, but it was a part of Henley. Anyway, Graham was giving full cover, that is to say 24/7, so it stood to reason that he would catch far more shouts than me. During the day, usually one of the two fire engines was OTR or "off the run". This was due to fewer than the minimum of eight men working in the town to crew both. So, Graham stood a very good chance of riding for every weekday daytime call. However, night time and weekends was different. With all of us home from work, it would become a race to the station. Now, this is where I cross my fingers and hide them behind my back when I say "We always completely obey the rules of the road and the speed limits when responding to the station for a shout". I will say no more on that, for the moment.

On my way home, that evening, I popped into the station to

see if they had had any shouts. There had been none " oh well a quiet day then", I thought. I got home, had some tea and then took the dogs out. We had two, one was an all black German Shepherd called Ria and the other was a golden long haired Dachshund called Rema. Whilst striding out over the field with them, as I had done ever since moving to Henley, about a quarter of a mile from home, the thought struck me, "what would I do if I got a call now?" I would have a hell of a run back to home and my car. Ria would keep up with me but poor little Rema wouldn't stand a chance and, more to the point, I would probably miss the pumps! This was my first proper realisation that this part time job was going to affect my life in so many different ways. Far more than I had thought about when I applied.

Again, that night I laid out my clothes along the floor and had a little try to see how fast I could get dressed. At between 20-30 seconds, not bad. I did have a little bit of warning for some shouts as, in those days, you could, if you knew the frequency - and surprise , surprise I did - listen in to the fire service radio. Nowadays, the old Oxfordshire frequency is BBC Radio 1. But then, you could listen to your neighbouring stations and hear what jobs they had and whether they put in an assistance message. When you heard that, you would get dressed and be halfway to the station before the alerters were operated. But, of course none of us had these radios or were sad enough to listen in. Again, my fingers are crossed!

Yet another night went by with nothing, and I trundled off to work. On my way home, again I popped into the station and there, in the log book, was a shout. I thought "Bugger, it had been during the afternoon and Graham had caught the pump, so his duck was broken. " It couldn't have been much of a job, as they were booked off duty in under 45 minutes.

When I got home, mum had tea on the table. It was my favourite, Steak and Kidney pudding. I sat down and, as I put my knife and fork through the thin suet outer, the rich dark brown gravy began to ooze out and the heavenly aromatic scent began to..... BEEP BEEEEEP, BEEP BEEEEEP, BEEP BEEEEEP, a shout! I jumped up from the table and, knocking over my glass of water in the process, ran to the door.

"Bugger, they weren't on the hook where they should be!"

"Where are my car keys Mum?" I shouted.

"Wherever you left them" she replied,

"Thanks" , I thought, sarcastically, "that was a great help."

I wasn't panicking but, then again, I wasn't exactly calm as I scanned the small hallway in our bungalow. We had a dark brown patterned carpet and, if you dropped anything on it, it wasn't exactly easy to find again quickly.

"Ria!", I shouted exasperatedly. She looked up at me, from her bed, startled, as if she had done something wrong, which of course she hadn't. There, under her huge black paw were my car keys. They must have fallen off the shelf by her bed in the hall. I grabbed them and ran out to my car, which I had parked up on the drive, already pointing out towards the road.

Because I had only just got home from work, the car roared into life immediately. Off I went on the route I had travelled hundreds of times but, this time, it was for real. First roundabout on our estate, left turn, no problems there. A couple of hundred yards later, the main road approached with another roundabout, this time a right turn. I slowed to almost a stop, which was frustrating, but the view was obstructed by an overgrown tree. Clear right, go. WHOOPS, I put my foot down a little bit too hard and the car slid just a little as I pirouetted around the bollards in the road. OK, now for the straight run down the hill in Greys Road to the town, remembering to slow down, I mean check my speed, as I passed the speed camera, flipping thing. Now at the bottom of the hill, hard left onto Paradise Road and up a hill again, my suspension groaning as I hit 4 speed bumps on the way, possibly not at the designed velocity for the road calming measures.

At the top of the hill, another main road, turn right by the entrance to George Harrison's mansion, yes of Beetle fame, then immediately left, right on to West Street where the station was.

However, you had to be extremely careful on this junction as at least 5 or 6 of the crew lived on the estate coming from the other direction and we all had to creep around the blind corner making sure it was clear. I was on the home straight now, a couple of hundred yards and I'll be on the forecourt. As I drew level with the Row Barge pub, I heard the two tones of the first pump turning out. "Hold back and let it go" I thought, " I don't want to have a crash on my first shout" . The shiny red nose of the Bedford eased out from the appliance bay and I could see John in the OIC's seat, looking up the road at me and giving the all clear to the driver.

As it pulled out onto the road, I slipped in behind it onto the forecourt. I jumped out of the car and locked it, then ran at full chat into the appliance bay. The first 6 tallies were missing from the board, the crew on the first pump would have taken those. But the second tallies were still hanging up. I grabbed one of the two green ones which I was allowed to take and parted the heavy plastic strips in the doorway to the muster bay. On emerging into the room, I got a sinking feeling as Bob and Tony were there but were not getting dressed. They were looking at the printout, It's a one pump shout. I immediately felt so deflated, my first shout, I had missed the pump. Disconsolately, I stuck my arm back through the flaps and hung the tally back on its hook.

As we waited, the rest of the crew arrived , seeing us through the plastic, they slowed down and just walked in. Now that we were on station, we had to wait for 20 minutes, just in case we were needed by the first pump. Bob went into the appliance bay and climbed into the cab of B6W to turn the radio on so that we could hear what they had. We sat around and talk turned to football and other similar topics to pass the time. Then we heard John's dulcet tone on the radio. It was a small bonfire which had spread to some trees. It turned out that it was at the actor George Cole's house. On hearing that, the old hands made their way to the pay sheets and signed off. I followed them and got into my car to head home to what would be my now somewhat deflated steak and kidney pudding.

"You're back quickly, weren't you needed?" mum said, as I shut the front door behind me.

"Just a one pump shout" I replied, trying not to show my disappointment at missing it.

"Your dinner's in the oven" she said handing me the oven gloves to retrieve it.

It was a nice meal, but it wasn't quite the same as when it had been fresh out of the oven. Little did I know, in the years to come, how many spoilt meals, including a few Christmas ones, I would either miss, or have to retrieve from the oven, dried and looking rather pitiful.

Friday was much the same as Thursday.It was an undisturbed night followed by another dreary day at work. Home for tea via the station to check the log book, no shouts, then back to the station in the evening for a drink and a few games of darts in the social club upstairs.

It was just my luck that, before I had passed out on the run, the station had been picking up at least one shout a day. But as soon as I was available they dried up. It's a strange feeling, which is still with me today, the waiting for a shout which would allow me to put into practice the skills I have. But at the same time, it also means that there is someone in trouble or distress. Of course, that is a strange juxtaposition in which to find yourself. This went on for over a week, with nothing coming in. The following Wednesday, a drill night, Graham and I took some ribbing about it being our fault that the shouts had dried up. Thursday and Friday were the same as the previous Thursday and Friday, nothing.

Saturday morning dawned bright and fresh and, just as the kettle went click for the first cuppa of the day, BEEP BEEEEEEEEP BEEP BEEEEEEEEEP!

I ran back into my bedroom to get dressed - back to the door, keys in the right place this time. Car covered in dew. I must get new wiper blades. The car starts second time with a bit of choke, and off I go again. The route is becoming second nature now except for the morning shopping traffic, which, as soon as I get some speed up, I get baulked, again and again.

I conduct the drive to the station with mixed feelings. Half of me was hoping that I would finally catch the pump, whilst the other half was seemingly resigned to the fact that it was going to be another single pump shout with me back home in just over 20 minutes.

Yet again, as I came down West Street, B06L is pulling out. Yet again, I pull in behind it. As I do, Graham appeared on his bicycle up the path at the end of the drill yard and darted across in front of me. I climbed out, locked the car and jogged towards the appliance bay. There was something different this time, there being no one milling about by the tally board. I looked through the plastic screens and saw the others picking up their kit. IT'S A TWO PUMP SHOUT YES! I looked at the tally board. Graham had already taken one of the green tallies, but the other was still hanging there. So I grabbed it and burst through the screen, narrowly missing Mick Goddard as he came through the other way.

I unhooked my helmet and slid my arm through the chin strap. I tucked my tunic over my arm and, grabbing the handles of my boots inside my leggings, made my way out to B06W. As I made the rather cumbersome climb up into the rear cab, Tony was already in the driver's seat with the engine revving. Bullet and Ivor were seated in the middle in the back, by the BA sets and Graham was on the far side. Mick climbed into the OIC's seat and, as Tony pulled out of the station called out, "it's only Borocourt".

Borocourt was, shall we say, a regular for the station. It was the area's mental health and secure unit. It had been a very large mansion building standing in acres and acres of land. It has long since been converted into luxury penthouse flats and houses but, then, regularly it took in overflow from Broadmoor and suchlike. It was situated right on the edge of our ground just outside Kingwood Common. It would take us around 15 minutes to reach. The call was to an alarm actuating and both of Henley's engines would be attending along with the one from Goring.

I must have been grinning like a Cheshire cat as we reached the bottom of West Street with Tony putting on the two tones to go around the town hall. My dream was coming true. To this day, still I prefer the sound of the air horn two tones, or nee naws as children like to call them. They have a lovely richness to them and are very directional. Almost instantly, you know from where an emergency vehicle is coming. The sirens of the modern era are too shrill for me and, because the sound travels so far and bounces off buildings much more easily, they are hard to locate quickly.

Nowadays, while working for West Midlands Fire Service, our trucks have both sirens and old fashioned two tones. This is for a very good reason. I have experienced this many a time. Usually, when responding from a two pump station to a call, depending on the skill of the drivers and the power of the engines, the appliances stay within 50 or so yards of each other. Nowadays, if using sirens, the driver is supposed to use the long wailing sound for long straight roads and the short "yelps" for obstructions and approaching junctions. The change in sound is supposed to alert drivers of our approach. This is all well and good if there is just one truck. Usually, the driver will hear the noise, glance in the mirror quickly and pull over to let the truck pass. However, if there are two trucks responding, very often the driver, on hearing the same wailers or mix of tones, will not realise there is a second truck so close behind. As soon as the first truck has gone past, the driver, without looking, will pull back out into the road and will frequently get a shock as they become closely acquainted with a large red bumper bearing down on them.

To avoid this, in some way we are now trained, if we are driving the second truck, to use the two tones rather than the wailers. This introduces a completely different noise from the first truck and gives the driver a better indication of a second engine approaching closely behind the first. Ocasionally, you get an idiot who either still doesn't look or has their radio up so loud that they can't hear the approaching danger. They are greeted with a blast from the bull horn, which usually does the trick!

We roared up Gravel Hill out of town. As we were bouncing about in the back, Graham and I were struggling to master the art of getting our fire kit on in the confines of the rear cab, as Tony was throwing the pump around the road. As I finished dressing I realised that Bullet and Ivor were just sitting chatting, their kit still on the floor, not attempting to put any of it on.

This puzzled me, we were on an emergency call and yet they were showing no sense of urgency at all. I spent the rest of the journey looking either out of the side window or forward through to the front cab watching how most vehicles moved out of our way. I say most, as there were some of them, and there are still many today 35 odd years later, whose drivers seem in a world of oblivion. Their car radios, or now ipods, are turned up way too high meaning that they cannot hear us approaching from behind. What's even more dangerous is that eventually, when they do eventually wake up to our presence they get the shocks of their lives. With their rear window full of a huge red monster, they panic and hit the brakes. I can't even attempt to count the number of times, as a driver, I have then had to take avoiding action, or acquaint the officer in charge to the benefits of anti locking braking.

As we disturbed the idyllic peace of Peppard Common and headed for Kingwood Common, Bullet and Ivor finally decided to get their kit on in a leisurely way. The experience of time served begins to dawn. They knew, to the second, how long the journey would take and were fully dressed just before we turned onto the half mile long drive to the building. This calmness would one day turn on them and bite me and them, hard on the bums.

As we pulled up at the front door John was already walking out and waved at us to turn around and head for home. It was a false alarm, as usual here! We remained in our kit all of the way home .

As we pulled up outside the station, Ivor climbed past me to unlock the side door of the station and then drag open the large red doors.

What happened next might seem a little simple, but it has stayed with me to this very day. We had only travelled 15 or so miles and had done no work as such but we washed the engine before it was reversed into the station. It wasn't really dirty, just a little dusty, but we washed it all the same.

This was the beginning of me, later in my career, calling myself "old school". The beginning of the instilling of very high standards and, yes, pride in my job and equipment. Every time those engines went out, we washed them on return. Admittedly, sometimes we washed only the bit on the bottom up to the top of the wheel arch, the bit which we called the skirt. However, if it was really dirty then the soap and brushes came out.

When I became the officer in charge of my own crew I began to install these standards or rituals in my fire-fighters. Sometimes, they would moan or pull my leg about it, but they still did it and took pride in their engine.

I feel saddened now that these standards are being lost. When I transferred to West Midlands Fire Service, I did so as a fire fighter not as an officer. One Sunday afternoon, on my station, most of the crew were in the TV room. I was a bit bored, so I hunted down a tin of red polish and started on the faded and slightly pink paintwork of our truck. Whilst in mid polish, a young new recruit fire fighter who was on standby with us that day wandered over.

"What ya doing?" he asked in a puzzled manner.

"I'm polishing the truck" I replied in a stating the obvious tone of voice.

"Why?" was his reply

"Because I take pride in my equipment and truck" I replied petulantly.

He just shrugged his shoulders and wandered off to the TV. I was saddened at his reaction, but was hit with the realisation that nowadays most people do not join, or see, the fire service as a long term career in the way I had those many years ago. I suppose that really, they use it either as a stepping stone or an entry in a CV in order to move onto other bigger and better things.

We finished cleaning and shamming the engine dry which, by the way, wasn't easy as I have previously described. The locker grills were not made of the smooth silver aluminium found on modern vehicles. They were a corrugated type painted red. You had to slide the leather along each groove, sometimes scrapping your finger along the uneven top edge.

I saw Tony back the engine into the station and got the mop to clean up the tyre marks.

As it's early in the morning I asked, "does anyone fancy a cuppa?" To that there were a few agreeable grunts, nods and "go on thens". I put the kettle on and, while waiting for it to boil, took off my kit and hung it up. The tea was not ideal as I had to use powdered milk, it not being worth keeping fresh milk on the station. Still, we downed our hot liquids and, once the other engine had been parked, we signed the log book and headed for our homes.

As I walked through the front door, mum just said,

"I take it you caught that one then?"

"Yep, first one down" I replied, a little smugly.

I didn't have to wait long for the second shout either.

It came later that day as I was sitting down with mum to watch the wrestling on World of Sport with Dickie Davies. Just as Big Daddy was about to flatten Kendo Nagasaki the alerter erupted into life again. Isn't it strange how quickly dogs pick up on things? As soon as the alerter sounded, both Ria and Rema ran for their beds and hit the deck. I think they had already cottoned on to the fact that, when they hear that noise, it's safer to stay low as a certain individual would be running through the house.

As I came down West Street, Bravo 6 Lima was on the forecourt but was not pulling off. Carefully, I pulled in front of it and parked. As I ran inside, I glanced across at the still stationary ladder and saw the lads getting dressed in the rear cab. Odd I thought, as I reached the tally board, I grabbed my now familiar green one and fought my way through the plastic into the muster bay. There were three or four others already in there, either just finishing getting dressed, or grabbing their kit and heading out to 6 Whiskey.

"What's the shout Mick?", I asked as I removed my kit from its peg.

"Regatta headquarters" came the reply as he disappeared into the appliance bay.

Now, I understood why 6 Lima was on the forecourt for so long. The new regatta HQ was on the far side of Henley bridge, less than 60 seconds from the station. It would look very unprofessional if the engine arrived with the crew still getting dressed. So, the first engine was held back to give the crew time to dress and get their BA sets on.

The Bravo 6 whiskey crew had a little more time to get sorted. As I climbed on, the others in the back were nearly dressed and we pulled off as I shut my door. I started to panic a little as I was now under pressure to be dressed before we arrived.

Helmet on, leaving the forecourt.

Tunic on, Velcro and buttons fastened, passing the town hall.

Boots on, passing Falaise Square.

Pull up leggings, half way down Hart street and approaching the bridge.

BUGGER!, braces.

In my rush to get dressed, I had completely mucked up. I had put on all of my top half clothing before my boots and leggings. Now, I had to undress, taking off my tunic to put on properly the braces of my leggings.

As we pulled up at the Regatta HQ, I was still only half dressed and, as I glanced to my left, the others all had wry smiles at my predicament.

They say, "what goes around comes around" and, more than 25 years later, I was reminded of that day. By then I was the officer in charge of my own retained crew in Banbury. We were turning out to a shout in the middle of the town, and I had a new recruit in the back, on his first call. He will remain nameless but, as I looked over my shoulder into the back to tell my crew what I wanted them to do on arriving at the scene, I saw that he had done exactly the same as I had done all those years before. I didn't have the heart to either bollock him or take the mickey. I just told them what I wanted and turned back to the front with a big grin on my face.

Now, I had broken my duck and had caught my first shouts however I had not even got off the engine when we were in attendance. I still didn't know how I would cope or react to a real incident when one came along.

On arriving at incidents, John had a rule, which usually worked.

"No one was to get off of the appliance until he had instructed them what to do."

One person, normally not a BA wearer, would get off with John and would follow him into the building or incident scene to act as his runner. In those days, we did not have personal radios and it was usually the newbie who became the link between the OIC and the pump. Fortunately, radios would be issued within a couple of months, and they would be a tremendous asset on my first real house fire. More of that later.

Over the next couple of weeks, the shouts trickled in, but there was nothing of much interest and certainly no flames. This was about to change.

My weekends were beginning to form a regular pattern. I would arrive home from work and have tea with mum. After freshening up and getting changed, around 8 o'clock, I would head off down to the station social club. In those days, most fire stations had their own social club and bar within the building. Some, like Henley's, were very successful and became meeting places not only for the firemen, but for their families and the retired firemen who used the club to stay in touch with their long-standing friends.

The Barmaid for the weekends was Tony's sister Pat. She was a jovial and sometimes larger than life character who brought to the place cheer and, often, gossip.

At the time, I was still single, with no females on the horizon. So, regularly, I found myself there in the evenings, playing darts, pool or cards, depending on who was there.

Now all the social clubs have closed down. The start of the rot began when the County Council, on all council properties, introduced a no smoking policy, followed by an alcohol ban, unless for specific events. This had the result of destroying the close-knit communities which thrived on most stations. People would splinter off to their favourite local and would sometimes not see one another until the next shout. I find this to be extremely sad and believe that it is one of the factors which eventually, and irrecoverably, turned the Retained Fire Service in Oxfordshire from a vocation into being merely a job.

CHAPTER 6
My First Goer

It was Saturday 2nd August, and I was about to baptised into the magical world of fire.

Apart from the social side of going down to the club at the weekends, there was a certain amount of, "bell hanging" associated with it. If I was at the club when the bells went down, barring getting trampled when racing down the stairs, I would catch the first pump. I would rarely achieve that when at home.

It is a bit of a misnomer talking about bells going down when, in fact, it was a loud siren type noise. Even today, old phrases from bygone eras are still coined.

"Bells going down" refers to the practice of ringing the church bells to summon fire volunteers to the fire house before the invention of telephones. Later, fire stations would have electrically operated bells in the appliance bays. They would start with a quiet ring and if that stopped then it would just mean that there was a message on the printer. If the ring didn't stop but burst into a much louder sound, then a shout was coming in. Before the advent of alerter pagers, retained firemen would have bells wired into their homes. During the hours of darkness, the sirens on the fire stations would not sound, so as not to wake up the neighbours. Often, I have heard the old hands say that the relays in the bells were so loud that they would hear the click of the switch and would be jumping out of bed well before the bell started ringing to signify that there was a shout coming in.

Here's another old term still in use today.

"A shout" This goes back even further to the times before the use of telephones and the 999 system and to when people would shout "FIRE" as they ran to the fire station. Even today, when carrying out a home fire safety check, we give the same advice. If the occupier is trapped upstairs then, shut the door and block the bottom with blankets, duvets etc to stop the smoke entering the room and get to a window and open it for fresh air. If they are going to shout for help, we tell them to shout, "HELP FIRE!", not just "HELP!". If people just hear the word help, then often they won't do so as it could be a mugging or something like that. However, if "HELP FIRE!" is shouted then you will be surprised how many people will come to help. I feel this is a little sad, but it's a reality of modern-day life.

Now at no time have I claimed to be a very good darts player.

I can hit a dart board in roughly the right area, but I cannot subtract quickly for love or money. My attempts are accompanied by under the breath tutting from both opponents and teammates. On this Saturday evening, I was plugging away at the board when the bells went down. simultaneously all three darts left my hand in the general direction of the board, as I turned towards the door. As I reached the bottom of the stairs, the delay in operating the pagers had been overcome and they also burst into life. The printer was still churning out its important information when I reached it to read,

B06L B06W

Barn Fire

Swiss Farm

Marlow Road

Henley.

"Both" and "Swiss Farm barn" I called out to the others who had downed their pints before following me down the stairs.

I left the printout on the printer as it was the responsibility of the officer or driver to tear it off.

I stuck my arm through the plastic flaps and, without looking, removed one of the two bottom green tallies.

Then it seemed as if everything went into slow motion. I had time to get properly dressed in the muster bay without elbows in my face and such like. Then I could walk to B06L and in a leisurely manner climb in the rear cab and just sit and wait, and wait, and wait for the others to arrive. It seemed an eternity until the first cars started to appear on the forecourt, but it was still just as fast as normal. It was just that I hadn't had to rush.

As we pulled out onto the forecourt, cars were slowing down and turning in behind us just as I had done many a time before. We drove through the centre of the town to many a cheer from drinkers outside the pubs on this balmy weekend evening. As we turned onto the Marlow Road, John called out from the front,

"It's a goer"

I leaned over to look through the middle of the cab to see a huge orange glow in the distance. My adrenalin started to pump. At last, this was it! As we drew nearer, there were people on the road waving us down. It seems strange, " Why are they waving to us?" Was it to show us where the fire was when it was blatantly obvious? As we pulled into the entrance, the barn was to our right, very close to the club house. Swiss Farm wasn't only a farm, it also had a large static caravan and touring caravan and camping site, with a large club house and bar.

As John climbed down from the cab, he was met by the owner who, in my opinion, was remarkably calm. There was a short exchange between him and John and we were all beckoned out of the truck. In a very calm but authoritative manner, John started issuing his orders. This was my first experience of being on the receiving end of orders at a fire and it has stuck with me to this day. In all my time in the service at Henley, I never saw John lose his cool or become flustered. This is much to his credit. This example has stayed with me through all my service, and I try to model myself on his example. They say that you never forget your first boss, for good reasons or otherwise.

For exceptionally good reasons, this is true of John. In fact, I am still in touch with him today, long after he retired from the Brigade.

John gave me my first task.

"Go and find the hydrant and set in. I think it's back on the road to the right of the entrance"

This was my chance to impress on my first working job and I was going to do just that. I ran off at full speed.

I got to the engine and opened the middle locker to get out the standpipe, key and bar, and then I was off and full pace back to the road.

First mistake, forgot to bring a torch!

Run back to the engine and get a torch out of the cab.

Start to run back to the entrance, stop again, take a length of hose with me as well!

Turn back to the truck again and get a length of 64mm hose out of the locker.

Now picture the scene if you will

I had a large metal standpipe with a key and bar stuffed under my right arm, had a roll of hose under my left arm, and had a torch hanging from one finger under the hose. And I was trying to run like this!

I must have looked comical to the crowds which were gathering from their mobile homes. It didn't cross my mind to ask any of them to give me a hand with carrying some hose. As I was doing this, I remember seeing the rest of the crew either walking or slowly jogging around to do their jobs. I remember thinking, in my naivety, that they could show a bit more enthusiasm.

I got out onto the road as B06W came down from the town and swung into the farm. I dropped all the kit I had on the path and started to look for the hydrant plate, nothing! I ran 50 yards towards the town, not a sausage! I went 50m yards the other way, again nothing! By then I was beginning to panic a little. The lads would have the hose run out and would be starting to put water on the fire, but the engine tanks wouldn't last that long, and I still hadn't found the hydrant. Again, I ran back along the path towards the town, and, in the dim beam of my torch, I caught a glimpse of a small manhole cover in the path. I stopped and looked closely at it and saw the letters FH on the lid, Fire Hydrant! Yes! found it. I looked around and, in the hugely overgrown verge, I saw the yellow top of the marker plate peeping out from the bushes. That's why I missed it the first time.

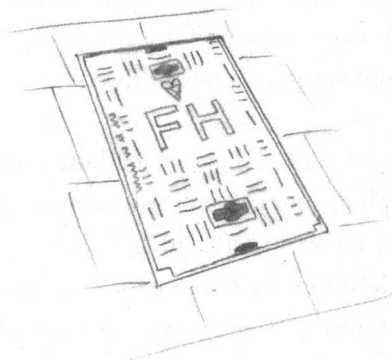

I ran back to where I had dropped the standpipe key and bar. I grabbed them and ran back to the hydrant. I pulled the bar out from the hollow tubular key and tried to use the wedged end to lift the hydrant cover out of the path. It rose up about half an inch and dropped down again. This happened about half a dozen times because the wedge on the bar was worn, and I couldn't get a good purchase on the lip of the lid. Eventually, I flipped it up and it landed upside down on the path with a satisfying clanging sound.

Staring into the black abyss that is the hydrant chamber, I realised that it was full of silt, and I stood no chance of fitting the standpipe unless I dug it out. I then did something which, nowadays, I would not dream of doing. I plunged my ungloved hand into the pit and frantically started scooping out the mud and silt.

Anyway, I finally got the pit sufficiently empty of the silt to fit the standpipe onto the top inch or so of the hydrant screw thread. Slowly, I turned on the hydrant and watched the dark brown muddy water gushing from the outlet. Once it had turned clear, I shut it off again and coupled the first length of hose to it. With sweat pouring from my forehead and the salt stinging my eyes, I began to run back toward the fire unravelling the hose as I went.

After about 20 yards I ran out of hose and dropped the aluminium coupling onto the path and continued running back to the engine for more hose. Running to meet me was Graham carrying a length of rolled up hose under each arm. As he drew level with me, I took one of the heavy rolls of hose from him and we both started running back to the coupling that I had dropped. He dropped his hose on the ground and connected it to the first length and then began to run back to the fire with his hose snaking behind him. As his ran out, I was behind him with mine ready to take over. He held out his left hand and grabbed my trailing end of hose as I passed him, snapping it into his coupling. We had to do this a couple of times more, running back and forwards to the engines for more hose. Finally, we reached the back of B06L and slammed the final coupling, with a satisfying double clunk as the lugs located on the rim of the collecting head.

Tony was the pump operator and, as we ran back again to the hydrant he shouted to us,

"WATER ON BLOODY SHARPISH!"

Graham stopped halfway back, where he could see both Tony on the pump and me on the hydrant.

When I reached the hydrant, whilst raising my right arm, I shouted to Graham at the top of my voice,

"WATER ON!"

He turned to Tony and again shouted.

"WATER ON!"

This was to confirm that Tony was indeed ready, which he was.

Graham turned back to me and replied

"WATER ON!",

With that, I turned the hydrant on, maybe a little too quickly in my haste, causing the hose to buck and jump like a cobra being rudely awakened from a sleep, as the water pressurised the hose and straightened out any kinks that were in it.

Next, because the hose had crossed the main road, we had to get the hose ramps off the roofs of the engines. These were heavy solid rubber ramps with two grooves for the hose to lay in as cars drove over them. I climbed onto the roof of Lima whilst Graham climbed onto Whiskey. As we threw them to the ground, we shouted

"STAND FROM UNDER!"

We were hoping that there was no one near because I, for one, certainty did not wait for a reply before launching the heavy ramp into the air. We got all six down and interlinked them to provide a wider target for drivers to aim at. Once they were in position, we put down blue flashing lights to mark them.

This whole episode had taken no more that 5-10 minutes, but it began to dawn on me why the others had, economically rather than slowly, been going about their jobs. I was on my chin strap breathing as if I had just done a marathon. I had wanted to impress John on my first job and had been doing everything at full speed instead of pacing myself. In contrast, to the others who had been pacing themselves and had energy in the bag should they need it.

Graham and I walked back to Tony at the rear of Lima. As we reached him, he quietly but somewhat sarcastically said,

"Both of you were wearing your high viz jackets whilst working on the road, weren't you?"

In other words, he had given us the heads up if John had seen us.

With that said, John did indeed come around the side of the engine.

"How are we doing for water Tone?" he enquired in his usual calm manner.

"Not too bad", was Tony's reply

"We're set into the hydrant but it not fantastic"

"Do you want it twined?" John asked.

Tony nodded, "It would be a help."

With that, I glanced at Graham. The look on his face must have mirrored mine and my heart sank.

"Off you go boys twin the hydrant please, quick as you can" John instructed us.

Twinning is running a second line of hose from a hydrant to a pump, basically repeating what we had just completed. The maths behind it is that when you double the diameter of hose, i.e. put in a second length, you quadruple the amount of water available to the pump. We both turned and went to get yet more hose. As we did so, I leaned into the rear cab and grabbed a couple of the high viz jackets and threw one to Graham. He caught it and gave a little nod and a grin.

"Good thinking lads", John said as he turned away from us and headed back to the fire.

Eventually, everything started to calm down and, after we had twinned the hydrant, I wandered round to the fire to see what was going on.

I was about to discover something very important. All the time I had been getting the water supply, I had been running at full speed whilst the others were slowly jogging about. Now I was absolutely knackered and needed a break, but the rest who had been pacing themselves were carrying on with their jobs, lesson learned.

On the branch, Rick was aiming the jet of water at the club house rather than at the barn. In fact, no one was putting water on the fire at all! Just then, John came up and in a puzzled tone, I asked him

"Station, why are we not putting the fire out?"

"The barn is already lost and the straw as well, even the unburned straw. As soon as hay or straw smells of smoke it is worthless to the farmer. Cattle will neither eat it nor settle down on it, so the best thing is to let it burn out under control" he explained very clearly.

"If we start putting water on the fire, all it will do is make it that much harder to turn over and it will smoulder on for days. Now, my greatest concern is protecting all the surrounding buildings and machinery."

As we were talking, one of the bar staff came out with a tray of teas and coffees.

"I'd rather have a pint", Rick quipped. This produced an exasperated sign from John.

Now that all the running about had stopped, the spectators began to dwindle and return to their nice warm beds while we were left to sit and watch this barn slowly burn down.

It was about 2 o'clock in the morning when our relief crews appeared from Goring and Watlington. While the officers were briefed, we took the hose off their engines and put it on ours to replace all the hose which was still in use. As I was doing this, I noticed Bullet acting in a cagey manner beside one of Goring's lockers. He then took quickly, but quietly, a dividing breech from the pump and put it onto ours.

"We haven't used a dividing breech Bullet" I said not too quietly.

"SHHHHHHH", he hissed at me,

"I know that, but we lost one a couple of weeks ago and I've been waiting for a chance to replace it" he whispered.

"But that's Goring's", I said rather stupidly

"I know that your berk. How else do you think we replace missing kit without heaps of paperwork and questions?"

"Oops, sorry" I replied sheepishly turning away to make sure that we weren't being watched.

The good thing about being relieved at the fire ground is that when you get back to the station you have very little equipment to clean as most of what you have is unused from the relieving engines. If you get the short straw and are the last to leave an incident, then you have heaps to wash and tidy up. We hosed down our boots and leggings, washing away all the burnt straw and ash.

While the others cleaned the engines, Graham and I carried out our now familiar task of making the tea.

By the time I crept through my front door, it was gone 3 in the morning.

As I tried hard not to trip over the dogs who had got up to welcome me home.

"Is that you Darran?"

came the faint call from mum's bedroom.

"Yes, sorry did I wake you?"

"No, I haven't been able to sleep" she replied tiredly.

"Why's that, are you not feeling well?"

"No, when you didn't come home, I began to imagine all sorts of things. I thought you might have had a crash and were lying in a ditch or something. When it got past midnight, I started phoning the hospitals to see if you were there. Eventually, I phoned the police station, and they told me that there was a big fire in town, so I knew where you were."

"Yeah, Swiss Farm" I said.

"I know that now but………."

"But what mum?"

"Look, your dad walked out the door and I never saw him again. When you go running out of that door, well, I don't want to lose you in the same way" she replied.

Ouch, it suddenly all became clear to me. The somewhat cool reaction to me getting into the brigade. The less than overjoyed response to me passing my test and becoming operational now all made sense. In my excitement, I had completely missed the fact that mum was really scared about something happening to me, and her losing me as well as dad.

"Don't be daft mum, I'm fine," I said, trying to brush it off and not make a big issue of it. However, to this day, I was careful with what I said about work to mum and, even more to my kids.

I had to have a shower as I stank of smoke and was covered in soot, I was MUCKY DIRTY GRIMEY SWEATY and loving it. That night, I slept like a log and didn't surface until mid morning. I had done it . I had got my first fire under my belt.

CHAPTER 7
Basic Training

The rest of the summer passed rather uneventfully with shouts coming in every day or so but nothing to really stress about. At the end of September, Graham and I took a driving test. This was so that we could use the brigade's vans because, in October, we were to go to Didcot for our weeklong basic training course. Also, we received our first probationary reports. We were called down to the station one Sunday Morning to meet with DO Collins and to hear what our Station Officer had to say about us.

To say I was rather pleased with my report would be putting it mildly.

It was much better than I ever got at school, when my history teacher Mrs Syrett, who I am still in touch with, called me "A lazy wretch of humanity."

(1) Performance on drills and training

Training started well, keen to learn with the gift of being able to retain all he is told. It's obvious this man spent time with his drill book, in the knowledge he's obtained outside training hours.

(2) Performance on fireground

Early days for this man with little fireground experience at this time, but as previously stated only needs to be told once, then he carries out his duties with confidence.

(3) Attitude to colleagues

Right attitude towards his colleagues, participates in all station activities on and off duty.

(4) Reaction to authority

Reacts well, and always with respect

(5) Personal appearance

Good, always a smart and clean appearance

(6) Was about course dates

(7) Observations (with reference to any outstanding qualities or weaknesses)

Pleased the way this man started his service. He's done well with all his training. The one problem this man must check is his excitement when the station receives a call. He's been made aware of this, and he must check it.

The DO then had to write his comments of the report which John had written.

A good report which augurs well for the future. Fm Gough has displayed keenness and enthusiasm in all aspects of his work. The need however to contain this over enthusiasm and replace with a more controlled approach when turning out has been explained to him. This I believe will be attained as he gains experience.

Let me now take a moment to recall exactly how this was explained to me. Picture the scene. I'm stood in front of the desk, in full uniform, in the office with the DO sat behind it, with John stood behind him. The DO asks me whether I'm happy with the report which has just been read out to me. I nod and reply "Yes sir."

He then asks, "Fireman Gough, have you heard the story about the old bull and the young bull?"

"No sir" I reply slightly puzzled.

"Let me tell you it," he says in a Jackanory type manner. "Well, you see there's this young bull and this old bull standing at the top of a hill. OK so far?"

"Sir."

"At the bottom of the hill is a herd of young heifers. Now the young bull says to the old bull—"

At this point, The DO speeds up his voice.

"'Cor, look at all those young heifers down there. Let's run down and shag one each?'"

The DO now slows his voice right down.

"'Nah,' says the old bull. 'Let's walk down and shag the bloody lot.'"

With this I glance at John who, stood behind the DO, has a wry smile on his face.

"Do you see what I'm getting at Fireman Gough? "The DO asks.

I smile back. "Yes sir."

I myself, have related this story to many of my new recruits in similar circumstances. I know that it may seem a bit like an old hand reminiscing, but it stuck with me and always will do so.

Much has changed in the Fire Service since I joined. One change is that basic recruitment and training is done together. When I joined, we had 16 hours of training on station, then, once that was passed, you were working for real. Then you had to attend training school within the first 6 months for a basic training course. This seemed a bit back to front. Nowadays new recruits have an induction course of four nights at HQ followed by a two-week basic course at training school before we ever see them on station. This was to even out and standardise the training which the new recruits received.

The rationale behind this was that when we arrived at Didcot on day one, we were all at different levels of knowledge and skills. That depended on the station we came from, the level of training from that station, and how busy that station was. The training school staff had to use day one and most of day two in assessing what we knew, or didn't know, and to try to get us all on a level playing field. Graham and I were among the lucky ones. That was because John had such high standards on station that we arrived very near to, or even, at the top of, the knowledge tree.

The major fault with the new system is that when the recruit finally lands on station, they are supposed to be ready to go straight on the run. In later years, as the Officer in Charge of my own crew, I had serious issues with this because recruits came to station knowing less after two weeks than I had done after one. Also, they didn't know any of their new crew mates and we had no idea of them or how they worked. As an OIC I would not, as a matter of course, put them on the run for at least three weeks. By then, I had sussed them and how much they actually knew, and I was happy that they were safe to put into dangerous situations. Whilst that was much to the annoyance of my station commanders, my response was "tough!"

I spent the weekend before training school down at the station going over the engine and the equipment and cleaning my kit. I had taken a week off work on annual leave, but the brigade paid for my loss of earnings, so I was quid's in.

Monday morning came and, at 0730, Graham and I met up in our best uniform at the station. That allowed us enough time to make the drive to Didcot for 0830. We loaded our kit into the van and decided to take turns at driving, me first.

As I turned onto the driveway beside the station and parked on the edge of the drill yard, there occurred the first of many full circles in my fire service career. Until I drove in, I hadn't realised that this was the station I had visited all those many years ago at primary school. Here, I was about to train on the very drill yard I had stood, and watched, in awe the crews demonstrating their skills.

We unloaded out fire kit and made our way to the back door. As we approached, a tee shirted fireman came out from the door and lit up a cigarette.

"You for the Training school?" he asked through a cloud of smoke.

We both nodded at the same time.

"Through here, dump your kit in the first room on the right. Then, then if you head upstairs, there's a brew on in the mess."

Again, we nodded simultaneously our thanks and then entered the station.

In the kit room we hung our fire kit on pegs alongside three other sets.

Then we explored the long corridor with numerous doors signed BA servicing and showers etc. At the end, there was a stone staircase which led to the first floor and the mess room. As we entered, we saw, at the mess table, the owners of the three sets of fire kit. They turned towards us and nodded a greeting.

"Are you here for the basic?" I asked.

"Yeah, we're from Banbury, where are you from?"

"Henley" Graham replied.

The beginning of two new circles was just starting. The first introduced himself as Chris Hirons. I would meet up with Chris again a couple of years later when he and I started our basic training as Fire Control Operators at HQ. The second was Brian Walker. Brian or "Shakey" as I would come to know him would still be in the Banbury crew when I moved there in 1992. I would spend many a year working with him. If the third one reads this, then I must apologise to him for forgetting his name.

"The pots just brewed" Chris said, so Graham and I headed over to the kitchen counter for a quick cuppa.

While we sat there chatting about our experiences and what we expected to happen that week, three more nervous looking bodies walked in, one of whom would become a good friend. We remain in touch to this day. Stuart Milsom was from B11 Wallingford, a neighbouring station to Henley. Over the years, we would meet up on many a job. We were awarded our 20 years long service medals together, and later we were awarded our Oxfordshire County Council 25 years long service award on the same evening. When I eventually retired from the retained service, Stuart became the only member of the October 86 course still operational with Oxfordshire. (However, when writing this book I am still operational with West Midlands.) Introductions over, we sat and waited.

0825 and a head poked itself through the door and shouted "ALL OF YOU OUT ON THE DRILL YARD, ON THE YELLOW LINE, IN FULL FIRE KIT IN FIVE MINUTES!"

With that, the head vanished and with a scraping of chairs, we all jumped up and followed the head out of the door.

Four minutes later, we were all stood on the same yellow line where I had stood with my Primary school.

Looking along the line, I could see that we weren't wearing the same kit. We all had yellow leggings but some had the Teled tunics. The three Banbury lads stood out like sore thumbs. They were black from head to foot. Their leggings did have some yellow peeping out through a black tar like substance. Their yellow helmets were in the same dilapidated condition. Only their tunics were clean. Before I had a chance to ask them why they were in such a state.

"CREW, CREW SHUN!"

We jumped to attention and a small, squat, dumpy, Sub Officer waddled out from behind us.

He tried to come to a smart halt, which failed, and turned towards us, it was the head from the door a few minutes ago.

The next of many, most of them unpleasant, circles was about to begin.

"MY NAME IS SUB OFFICER HEMMINGS, AND WHICH EVER WAY YOU SPELL IT, IT COMES OUT BASTARD!" the head below the helmet bellowed.

No, it comes out WANKER, was my immediate thought. Our relationship went downhill from there.

Our paths in the fire service would cross many, many times. Most of them would not be pleasant.

As he scanned the line, his eyes stopped at the three from Banbury.

"WHAT THE BLOODY HELL DO YOU THREE THINK YOU ARE DOING TURNING UP IN THAT STATE?" he bellowed.

Chris had the courage, or was foolhardy enough, to answer first.

"We had a bitumen tanker on fire last night and couldn't get any replacement kit in time Sub"

"IT'S NOT SUB, IT'S SIR TO YOU" was the voluminous reply followed by

"AND WHAT THE HELL ARE YOU DOING RESPONDING TO SHOUTS SO CLOSE TO THIS COURSE?"

"If we didn't, then the pump wouldn't have turned out SIR" with emphasis on the SIR, came Chris's argumentative reply.

"I DON'T CARE. YOU WILL NOT RESPOND TO ANY CALLS WHILST ON THIS WEEK'S COURSE. DO YOU UNDERSTAND?"

Chris answered back argumentatively, "But that will take the pump off the run SIR" with even more emphasis on the SIR

"I DON'T CARE. THAT'S AN ORDER" he bellowed again.

I began to wonder if he had any other volume than loud. It was also becoming clear that he was the sort of person who was always right, and it would be a waste of breath to argue with him, ever.

With the volume slightly diminishing, as he had made his mark, he continued,

"You will phone stores during tea break and arrange to collect fresh kit on your way back to station tonight. Understood?"

"SIR", the three of them replied in chorus.

We then spent the next hour being shown around the training school appliances and what equipment they carried. They were basically the same as the ones which we had on station, but some kit was either older or not required. The rear cab doors were of the folding type. To mount them you had to thump the middle of the door which then folded inwards away from you. With the tour over, we had twenty minutes to spare before tea break, so we were detailed to do some hose running. Some would say it was to see at what level we were in our training. Others, including me, would be a bit more cynical. On the first morning of day one, we were already victims of time filling.

Running out hose then rolling it up, then running it out again and then making it up again and so on for the spare twenty minutes.

Morning exercise over and up for a cuppa. On our way up Chris diverted into the training office to phone stores. When he finally reappeared, he seemed exhausted.

"Stores wouldn't take my word for it, and I had to get Hemmings to take over the call. I wouldn't want to be on the other end of that phone. Anyway, they will be at reception on our way home, but we've got to have the paperwork completed and he's reluctantly doing that now."

After tea, we were told to be ready in the classroom for 1100. On the dot of 1100, another Sub Officer walked in, and the phrase chalk and cheese springs to mind.

"Morning all, I'm Sub Officer Sadler, but in here it's Mick."

He continued to introduce himself, giving us a quick resume of his career and experience to date. He had over 20 years in and had seen a lot of changes, some of which we would find out about during his lessons.

Mick was the sort of person whom you automatically listened to and would follow. Yet another circle just started because his wife Ann would become my boss in fire control in a couple of years' time.

Mick started with listing all that we would be covering during the week, from pumping theory and fault finding, to combination drills involving multiple fire engines. There would be a practical assessment and a theory exam on the Friday, and they would be graded pass or fail. While he was talking, the Station Officer in charge came in and Mick called us to attention. As we all jumped up, Station Officer Moors waved his hands and said,

"Sit, sit, sit, I work for a living."

With that, we relaxed.

"Sorry to interrupt Mick. I just wanted to say hello and to introduce myself. I'll be popping in and out throughout the week and will give the Subs an occasional break and take a session or two. If you have any problems or issues, please feel free to come and see me any time, OK?"

We all nodded in agreement.

"Fine, I'll see you all a bit later, and more importantly do enjoy this week."

With that, he quickly left the room.

Afterwards we ordered our lunch and began a session on the theory of basic hose and pump drills. It felt just like being back at school but, this time, I was lapping it up. All we were being taught felt like it had a purpose, and we would be using this information in the real world. The one subject which did ring true with my schoolwork was the maths involved in firefighting. The calculations and formulas involved in the estimation of pressures and volumes of water soon came back to me, and I quietly thanked my inspirational maths teacher, Mr Howland, for not giving up on me and helping me and my friends with free private lessons because he left our school a few months before our exams were due.

Time flew by and, before we knew it, it was lunch. The food was really good and, afterwards, we collapsed in the station social club, chatting about the morning we had just experienced.

Subconsciously, we must all have looked at the clock at the same time, because we all got up in unison and headed for the kit room to change. 1359 and we were back on the squad line waiting for our instructors to arrive. At 1400 Hemmings marched out again but this time with Mick Sadler walking behind him.

Before Hemmings had a chance to bark out anything, Mick asked us if we had all had a good lunch, to which we all replied at top volume "YES SIR". Mick seemed a little taken aback by our forceful reply, but I think he realised we were playing up to his partner in crime.

We were to carry out basic ladder drills that afternoon and were split into crews of four. Even though I had "thrown" many a ladder up, and knew what to do, I was still a bit nervous. When you train all the time with your own crew you get to know each other's foibles and quirky ways. When attending a large incident, it is always best to remain as crews and not to mix if possible.

Hemmings detailed the drill, and I ended up as the OIC. On the "Get to work" the four of us ran to the appliance, slipped the ladder off the roof of the fire engine and headed for the tower. All good so far. The ladder was sighted and pitched to the correct window and on the side required. Bang, in it went. Spot on, happy bunny.

Then we were all told to climb to the top of the ladder and to dismount into the tower. That left Graham and me at the bottom, footing the ladder. As the first of the crew started to climb back down the ladder, I prepared myself to shout the hazard warning orders. The ladder has a couple of pulleys and pawls to be aware of and you have to move your feet accordingly. The correct call is "PAWLS.........STEP IN" This is called when the climber is one round above the hazard. At Henley, the crews had put in an extra call of "PULLEYS...... PAWLS........STEP IN" Not knowing that this was a Henleyism, I shouted the call to which I was used to. Oh boy, how I wish I hadn't.

As soon as I had done so, Hemmings erupted.

"WHAT THE HELL WAS THAT GOUGH? THE CORRECT ORDER IS PAWLS......STEP IN. IS THAT UNDERSTOOD?"

"YES SIR!" I shouted back knowing dammed well that it was not worth wasting my breath trying to explain that this was the order which I had been taught.

I glanced around and was met with looks of pity and relief from the others that it wasn't them on the receiving end. The rest of the afternoon was taken up with more and different ladder drills broken up with a quick cuppa halfway through.

At the end of the day Graham and I jumped into our van and

headed home whilst the Banbury lads were off to HQ for their new leggings.

When we got back to the station, the lads were just back off a shout and were washing the pumps off before putting them away. We were asked how our first day had gone and who our instructors were.

When I said Hemmings, it was met with raised eyebrows and blowing out of breath together with "he's a right......" to which I had to agree.

However, we got a different reaction when we said Mick Sadler and Henry Moors. Both Mick and Henry were deemed to be OK. But more than one said of Henry, that if you want an easy day then start him off talking about sailing. He is an avid sailor and can talk for hours about it.

When I got home, my tea was already on the table. After tea I went for a long soak as the afternoon's ladder pitching was beginning to take its toll. It felt strangely reassuring that I could go to bed that night and for the rest of the week in the knowledge that I would not be disturbed. This was because we had been ordered not to respond to fire calls whilst on the course, so I had taken the batteries out of my alerter. With that, I drifted off into a very deep and refreshing sleep.

The following day, it was Graham's turn to drive and we arrived early enough again to grab a cuppa in the mess before the day started. The Banbury lads had, after a palaver of chasing halfway around HQ, got their new leggings. At 0825 the door opened, and we braced ourselves for a Hemmings early morning blast but, to our surprise, we got the Mick Sadler, Dads Army Sergeant Wilson type greeting.

"Morning lads. All, OK?"

"Yes thanks, Sir" came our enthusiastic chorus.

"I told you with no one around it's Mick" came the acknowledgement.

"Anyway, five minutes and in the lecture room, the Station Officer is taking you this morning for message procedure"

"OK Mick" came the now rather slightly subdued reply.

0830 and we were in the lecture room with pens, pencils and paper at the ready. Stn O Mears opened the door and before we got a chance to stand up, he said "sit, sit, sit" along with a nonchalant wave of his arm. Henry reminded me a bit of my dad. He was in his 50s and had grey hair and a weathered face. We passed a few pleasantries about the previous day, the food and such like. He then told us what the lesson is about and raised his voice when talking about things which we might need to remember for Friday. With that he gave a Monty Python style "wink, wink."

He began with explaining to us the radio scheme and how it works, the frequencies etc. A wicked thought entered my mind and I put my hand up,

"Station?" I said in an inquisitive tone.

"Yes, Fireman Gough what is it?"

"I was wondering whether the radio scheme is similar to other emergency services or to someone like the coastguard for instance."

"Funny you should say that, yes, it is. I have a boat on the south coast and ...."

"Got him!", I thought, and I glanced at Graham who grinned at me as he also remembered the tip off which we had from the lads at Henley. We then spent the next twenty minutes or so listening to his sailing exploits.

Eventually, we got back on track and what was supposed to be an hour lesson on radio procedure was crammed into thirty minutes.

"Go and have a ten-minute break and then meet the Subs on the yard in full kit," Henry instructed as he waved us out of the room. While we were grabbing a glass of water in the kitchen the rest of the lads were asking me and Graham what all of that was about? We let them into our little secret and, for the rest of the week, whenever we had Henry to teach us, a different person would drop a naval question into the subject. Off again he would go, much to our amusement.

We spent the rest of the morning learning about the Lightweight Portable Pump (LPP). Most of this was a refresher to Graham and me but there were some nuggets of information and tricks of the trade which we picked up. Little was I to know that the training I received on that day would come in very handy in a couple of weeks time.

The rest of the week really followed the same pattern. First thing we would have either Hemmings shouting at us or Mick talking to us rather smoothly. That was interspersed with drills in the yard using either ladders or pumps. We would be split into two crews of four and each being assigned a pump on which to ride. We would then be "Turned out" from the front of the station to drive around to the yard and receive our instructions for an imaginary emergency.

One morning, we were taken through the procedures for dealing with a car crash and how to assemble the hydraulic equipment. This was really useful because "on station" the few RTAs which we had attended had Graham and me feeling a bit like spare parts. The old hands always got stuck in and left us to do the more mundane tasks of conning the scene off and providing lighting rather than using the "big boys tools".

All too soon, Thursday afternoon came around and we had a revision session to prepare for the written exam on Friday Morning. Mick took the session, and we were firing questions at him left right and centre. He had a very clever way of letting us know what was sensible to learn and remember and what was not. Every time we hit on something important for Friday he would say,

"That might come in useful to know."

With this, we could tailor our work for the morning.

That evening, on the drive home Graham and I fired questions at each other the whole journey. When I got home, I had tea and then spent a couple of hours going through my notes and the drill book for the following day.

Friday arrived and, as on the previous evening, we played Mastermind on the journey to Didcot. All of the lads arrived early and we gathered in the mess. Over an early morning cuppa we chewed the fat and questioned each other over bits and pieces. 0825 and Hemmings stuck his head through the door and talked to us, rather than shouted at us. They were running a bit late, and we had an extra half hour before the exam. So, we put another pot on and chewed the fat. Some of us questioned each other while a few just went over to the comfy chairs and read their notes quietly to themselves.

Soon enough, Hemmings appeared, and we put our notes away and followed him down to the lecture room. As we entered, the layout had changed with all the tables and chairs turned around so that no one position overlooked another. We all selected a table and sat down casting furtive glances and grins at each other. On the tables was small stack of paper face down with a couple of pens and pencils alongside.

Mick then came in with a "Morning lads".

We were told that we had one hour to complete the papers. They were divided into various topics. Some were multiple choice and the rest were short answers, with emphasis on the word short.

Hemmings looked at the clock and it was now ten past nine. "Ok, off you go. It finishes at ten past ten, but you won't or shouldn't need all that time"

With a degree of nervousness and excitement, I turned the paper over. On the cover page I entered my name, rank and station.

Immediately, I knew the answers to the first couple of questions. That calmed me down and gave me the confidence in myself to carry on through the paper.

I glanced up to see all of the others with their heads down, scribbling and crossing away. All was quiet except for the ticking of the clock.

After about half an hour, I was on the last question. That concerned me a little. I thought that I had gone through it too quickly. I rechecked my answers a second time. There were a couple of questions which I had to guess at but, on the whole, I was happy. I wasn't the first to leave the room but I breathed a sigh of relief as I handed my paper over to Mick and headed to the mess for a well earned cuppa.

The rest of the morning was spent chatting about incidents and the like, interrupted by a flying visit from DO Oliver, the senior training manager. Lunch was consumed in a somewhat nervous atmosphere with the afternoons practical drill session looming.

At 1400, we were all sat smartly to attention in the lecture room when Hemmings and Mick came in. After the usual pleasantries, Hemmings got down to the afternoon briefing. We were to be split up into two crews of four, one crew for each appliance. The scenario was to be a house fire with persons reported. We were to respond from the front of the station and treat the session as a "WYSIWYG"(What You See Is What You Get).

"Oh, and one more thing" Hemmings casually threw in.

"Fireman Gough, you will be the officer in charge."

Oh Great! Not that I wasn't nervous enough about the afternoon, but now I had to tell everyone else what to do.

We headed down to the kit room and the chat turned to who's going to do what? We started to plan as we didn't want to show ourselves up and, more importantly, we didn't want fail and be back coursed. It all began to get rather complicated with "I'll do this, and you do that". I had to call a halt to that. My thought went back to John and his rule that no one gets off the engine except for him plus one other. So, I nominated Graham to follow me at the start and the rest to await their orders. Everyone seemed to be in agreement, so we mounted the engines and waited to be called out.

The dulcet tones of Hemmings echoed from the yard behind us and off we went. As we pulled into the yard, I hit the two tones, why not? We pulled up to the side of the tower and I climbed out of the cab as Graham followed me from the second pump. Mick was playing the homeowner and told me there was a fire on the second floor and that his mate was missing. I quizzed Mick a bit more as to what caused it etc and then beckoned the crew to me. I briefed them quickly on what we had and detailed the crews to ladder pitching, hose running and hydrant setting in.

Everyone ran off and, suddenly, I felt all alone in the middle of the yard with Mick. Hemmings was standing by the station with his new toy.

A video camera, the size of a small suitcase, was perched on his shoulder and ominously aimed at me. Mick turned to me.

"Darran what are your thoughts now?"

I really wanted to say totally numb but thought better of it.

"Radio message to control confirming persons reported and possibly a make up" (request more fire engines in the real world)

"Good, how many pumps and what about an ambulance?

"Make pumps four. Do I need to ask for an ambulance as don't they get informed by fire control at time of call?"

"They do, well done".

Then I wandered off, looking purposeful, of course.

The lads had the ladder pitch bang on and the hose was running out nicely. I wanted to get involved but knew that would be a mistake.

The first lad went up and through the window and immediately shouted out "CASUALTY" With that, I shouted to send someone up the ladder to help with the casualty. The legs of the dummy appeared out of the window and Brian, who had gone up the ladder, secured the dummy between him and the ladder, and walked it down. On reaching the ground, we took the casualty from him and put it behind the tower.

With that done, the water was got to work into the tower.

Mick called me over and quietly said

"Knock off, make-up".

I acknowledged the order and turned to the crews. With both pumps running at full revs I shouted,

"KNOCK OFF MAKE UP" and held my hands out to my side and dropped them.

Everyone else shouted back "KNOCK OFF MAKE UP" and the pumps went silent.

With that, I went to help make up the hose and put the ladder away.

I only took one pace when echoing across the yard came, "FIREMAN GOUGH WHERE ARE YOU GOING?"

Hemmings was making his way over to me with the suitcase still aimed at me.

"I will now be a TV reporter and you are about to be interviewed."

Oh great, I thought that's all I need now.

"YES SIR"

We moved over to the side of the yard, and he started questioning me.

"Well Mr Gough what were you met with here and what were your actions please?"

I began describing what Mick had said when we arrived and what the crew did.

"A casualty was discovered and rescued but was found to be deceased "I continued.

Hemmings face lit up.

"When did you qualify as a doctor Mr Gough?"

Oh Shit! I thought.

"I'm not sir," I replied shakily.

"Then what gives you the right to declare someone dead Mr Gough?"

I remained silent. He put the camera down and, with a look of disgust on his face, told me to go and help the others with the make up.

As they had finished making up the kit, the lads were asking me how the interview went.

"I've no doubt you will see in a bit" was all I would say.

I kept asking myself what the hell was the purpose of it. As a new fireman, I would never be speaking with a TV crew, and what was its purpose in basic training?

We put the pumps back in the appliance bay and got changed. As we were heading to the mess, Mick called out to us.

"Stay in the mess lads and we will call you out one at a time."

This was met with grunts and nods as we disappeared into the mess.

One by one, we were called out but didn't return before the next was called out. Then came my turn. I walked smartly into the office and was met by Mick on the other side of the desk. I was told to sit down and then Mick passed over a piece of paper. On it were the marks for written examination 50 out of 50 and practical 35.8 out of 50. Total pass of 85.8, happy days.

There were comments from Stn O Moors and DO Oliver which were all very positive but the thing that sticks in my mind was something that Mick said to me.

"Darran you are exactly the sort of person we need in the brigade as wholetime, you should seriously consider it."

Here started another circle. It took me 23 years to finally get in the operational whole time but at my leaving do from Oxfordshire, Mick, who had long since retired, turned up and we both smiled at each other, remembering that moment in Didcot so many years ago.

When I left the room, I was directed to the lecture room where the others were. When we were all together again, Mick, Henry and Hemmings all came in for a debrief on the course and for any suggestions to improve it.

Then to finish, they ran the video from the afternoon. This

was all well and good, except that I knew what was coming. Sure enough, there was an eruption of laughter when I was interviewed. I looked a right idiot on the film, especially when asked when I had qualified as a doctor. Hemmings looked at me and asked for my comments. Thinking it best to keep silent to avoid dropping myself right in it, I merely shook my head.

Hemmings came up with some holier than thou comment which I honestly cannot remember as, by that time, I had switched off.

On the journey home, Graham made some supportive comments, but I was just glad to pass and hoped that I wouldn't see Hemmings again. Oh, how wrong I was!

CHAPTER 8
Earning My Spurs

It is said that, in everybody's life, you have at least one life changing moment. Mine came on the Monday following basic training week. I drove into work as usual and, after making the first round of teas, was quizzed by my employers, Ray and Paula, as to how the week had gone. With de-briefing finished, I picked up my work for the day. This was larger than usual as it also included some overflow from the previous week.

Once again, I found myself sitting in the darkroom with the red light on. But this time, instead of reading my drill book, I was just listening to the radio and thinking about the week which had just gone. I had lost all track of time when I was suddenly snapped back to reality.

"Tea's up Darran," came the call from Ray, who was stood outside the locked door. Already, it was eleven o'clock

"Thanks, just leave it outside, I'll be out in a minute." I couldn't let him see that I had done absolutely nothing at all that morning. I stuck my hand out of the door, found the mug and pulled it back in. I thought that I'd better do some work to show for the day but still found my mind wandering back to Didcot.

That day, I made the decision that, having spent a full week being a fireman, the photographic world was no longer for me. I would start looking for a job in Henley. That would allow me to respond to the station giving 24-hour cover, whilst I began applying to any brigade to become a fulltime fireman. With that, I began scouring the local rag for anything appropriate for me in town.

Wednesday night and the communications officer visited the station with "presents" for us. That day, two battery chargers had been fitted on each of the pumps and he came with four portable radios for them. Up until then, John or Mick had taken a fireman with them if they were going to be some distance from the pump. That would facilitate the officers getting messages and orders back to the crew or driver to radio fire control. This job of being the runner usually fell to the newest members of the crew.........me and Graham, at that time.

The radios had a range of about half a mile, which would be enough for our purpose. Little did we know that within a couple of weeks we would be singing their praises.

Yet again, I found myself wiping the sleep from my eyes on a Sunday morning whilst enroute to the station. As usual, B06L is pulling out as I pulled in behind it onto the forecourt. As I was getting my kit on Bob, who was already in the driver's seat of B06W, called out,

"Control on the radio, repeat calls being received they're making up" (adding more fire engines to the mobilisation before the first engine is in attendance. Usually due to the number of 999 calls being received in fire control)

As I climbed in the rear cab, Mick called out from the front,

"Ferry Cottage Lower Shiplake roof fire"

My old stomping ground where I grew up. I knew it well. With cab doors slamming we headed off to Shiplake. While we were still five minutes away, we heard John booking in attendance on the main radio with control and immediately making pumps 8. I thought "we've got a job on here".

As we turned onto Station Road, John called up on the new hand-held radios.

"Mick, are you receiving over?"

"Yes John, over."

"Mick, we've got a roof well alight, but we can't get the pump anywhere near the house. We will have to walk across a field for at least a quarter of a mile and we won't be able to get the kit there over"

"All received John, what do you want us to do over?"

"Mick, as you go over the level crossing, don't turn right onto the lane that runs alongside the platform, received?"

"Yes, received"

"Go down Bolney Lane and take the first right down to the old boat yard. There is a boat on its way to you there. Get the LPP and as much kit on it as possible, over"

"All received John, wilco"

As we near the railway station, I saw a sickening huge pall of smoke in the distance. Bob took the level crossing at, shall I say, slightly too fast a speed. All loose kit in the cab took off. We swung off Bolney Lane and down the unmade track to the boatyard which has since been converted into a very nice house. As we pulled up, between us and the river there was an area of grass which was about 25 feet wide. We saw a boat coming down the river towards us, so Bob pulled onto the grass and parked near to the riverbank. This was a simple and straight forward action which, in the days to follow, was to cause almost as much trouble as did the house on fire.

The boat coming to help us was from the Wargrave sailing club which was about half a mile upstream from us. It was a powered wooden dingy which was used to place the buoys in the river to mark out racing courses for the boats. That wouldn't be happening just now, as the river was in full flood following a

The boat was approaching us at an extreme rate of knots and, as it drew level with us, the man on board swung it around to face up stream. Now, it came to a virtual standstill, partly because of the fast flow of the river and partly because we were on the outside of a bend in the river, the fastest part of the stream.

The engine was on maximum revs and was crawling only very slowly back up stream towards where we had parked. When he got within reach, the man threw a line to us, and we pulled him in and tied the line to a fence post.

From where we were, we could clearly see the roof on fire and surrounded by weeping willows. Whilst it was only a hundred yards or so from us, we had to reach it by boat. (Health and safety breech number one, using a non-fire service boat which had not been risk assessed).

Mick explained to the boat owner what kit we would need to get on board. He confirmed that the boat could take it. Four of us lifted the lightweight portable pump from its cradle and manhandled it to the water's edge. At this stage I started to get a little concerned. (H&S breech number two. Crews not being tethered to a point of safety when working within three meters of the water's edge). Even though we had tied the boat to the shore, it was jostling and bouncing about like a child's balloon on a piece of string in the wind. The river was erupting and frothing up through the gap between the boat and the bank. I had this image of us trying to lift the pump over the gap between the boat and bank parting, resulting with the pump falling into the river. Graham and Tony climbed into the boat and nearly straight out the other side into the river as it rocked with their extra weight. The owner grabbed Graham, who in turn grabbed Tony. They steadied themselves. (H&S breech number three, crews not wearing lifejackets). Bob and I held both handles at either end of the Lightweight Portable Pump, (LPP), (Health and safety Breech number four, the LPP is a four person lift not a two). Struggling with the weight we half slid, and half lifted the pump over the frothing gap between the bank and the boat. Bear in mind the fact that the river was in flood, there was no drop onto the boat, the water being level with the top of the bank, so we had to lift onto the boat. As they pulled the pump on board, the boat dramatically dropped lower in the water. With the pump precariously balanced on the middle bench seat, Bob and I started passing the hard suction hose over onto the boat. Imagine a room size rolled Turkish rug but with double that weight. We were passing four of these huge hoses over onto the boat and watching it sinking lower into the swollen river. Next, we threw on as much soft firefighting hose as possible together with the branches, (nozzles) to go in the ends. This was followed by bags of rope, suction wrenches and cans of petrol for the pump.

The gunnels, (edges of the boat) were now perilously close to the river.

"On you get Darran," came the instruction from behind!

"Oh great", I thought as I gingerly stepped onto the already seriously overloaded craft. (H&S breech number five, overloading)

Bob untied the rope from the fence and, as we gingerly pulled out into the stream, the boat began, unnervingly, to roll to the left, or I should say to port. I leaned instinctively the opposite way and nearly too far over the edge. We eventually established equilibrium, of sorts, and set off toward the smoke.

I remember looking down at the fast flowing; murky waters and my thoughts went back to that awful Sunday when I lost my dad to this very river. I was now putting myself in a similarly dangerous situation. If we capsized then, with the weight of our fire kit, we wouldn't stand a hope in hell of surviving. I thought that if my dad could see me now and what I was doing, and the stupidity of the risks we were taking on this extremely dangerous river, which had already claimed his life, then he would have an absolute fit.

The "PUTT, PUTT, PUTT" of the engine was now straining to make any headway at all against the river which was roaring past just a couple of inches below us. We were crawling towards the house, at no more than a painfully very slow walking pace.

I could see John stood at the water's edge looking down stream at us and watching the dangerously slow progress towards him. Eventually, after at least ten minutes, we had travelled the hundred yards and finally drew level with him. Where he stood there was a slight indentation in the bank which created both an eddy and an area of slightly calmer water for us to pull into. I threw John the line and he pulled us tight into the bank. I have never been so relieved to get off a boat as I was then. The strange serenity of the boat ride was now blown apart with the frantic actions which were then exploding on the riverside. Between the four of us, we pulled all the hose and hard suction off the boat. I am sure that I heard the brave little tugboat breathe a sigh of relief as it once again rose up in the water to a safer level. Finally, the LPP was half heaved and half dragged onto the riverbank.

There were still only the four of us at the house as the rest of the crew were either back at the boathouse or making their way on foot across the water-sodden fields from the railway station to the house.

John was not a stuffy officer who just stood back and told you what to do, as was probably expected of him by the senior brigade management. No, he grabbed the hose and branches and started running them out toward the house. Meanwhile, Tony, Graham and I began assembling the hard suction hose and connecting it to the pump. When this was nearly complete, Tony shouted to me to get the pump started.

"OK, let's get this right!" I thought to myself as I flicked the on off switch to the on position.

Check!

Fuel supply turned on.

Check!

Chock on.

Check!

Right, here we go! Usually, this pump took three or four pulls on the rip cord to get it running and these were heavy pulls, at the best of times. I suppose it was because of the adrenalin created with the incident, which you don't experience on training, but I found myself in complete and utter shock at having pulled it so hard that it started on the first pull. For a couple of seconds, I was stunned at my strength but was brought rudely back to reality because the engine started to splutter. "NO, you are not going to stall on me now," I thought as I eased the chock shut and breathed a sigh of relief as the engine tone settled.

I dropped the hard suction into the river and turned back to the pump. Graham and Tony had left me now and were running after John to take over his hose and to run out a second. I could see they weren't quite ready for water which gave me time to prime the pump. I opened the throttle to full revs and pulled the priming lever down. Standing with my feet tight against the sides of the suction, I listened for the drop in tone of the engine to signify that the water was being forced up the hose to the pump. This was followed by feeling the vibrating water through my boots as I gripped the hose with my feet. Finally, the engine's exhaust turned to steam as the pump primed. Slowly, I opened a delivery and water started to pour out under pressure but suddenly dropped to a dribble. I began to worry but before I could do anything it burst back up to full pressure. Happy days! As I shut down the valve, Tony shouted,

"WATER ON DAZ!"

No pressure or branch number as I was used to in training, but this was for real. I could see him and could slowly open the delivery whilst watching him to make sure that he could control it. This was the real world, the world I had been training for. Just as he had water on and was training his branch on the roof, Graham hollered,

"WATER ON DAZ!"

Again, slowly I opened the second delivery and watched him tense as the water reached him and he started to hit the roof with his jet. Then I opened fully both valves and watched the gauges, compound, revs and oil temp. All were spot on the money.

Then I started to catch my breath and looked around for any housekeeping which I needed to do. The hard suction in the river needed pulling around to face the flow which was no easy task. Eventually, I managed it and tied it off to a willow tree on the bank.

I had to stay with the pump and monitor the pressures and make sure that Tony and Graham had all they needed. Then I found myself with time to look around and take stock of what was happening. The four of us had been there for about ten minutes on our own when I heard two tones in the distance. Our back up was arriving. Then the plucky little wooden boat returned from another run to the boat house with more crew and equipment on board. This broke yet more H&S rules. In the distance, on the far side of the house, I could see the gathering crews trudging over the water- sodden fields carrying more kit.

As the crews arrived on scene, we could start fire fighting for real. With only the four of us, it had been more a PR exercise, as we couldn't do anything substantial regarding putting the fire out. We could only hope to hold its progress in consuming the house until reinforcements arrived.

Rick came around and found me.

"I'll take over the pump if you want me to?" he said.

Translation.

"There is going to be a lot of hard work coming and I'll take over a plum job to watch the pump whilst the young'un goes and does all of the hard work. "

In all honesty, I wasn't happy as I was already getting bored seeing people getting stuck in whilst I was just stood there. Don't get me wrong, apart from BA entry control officer, the pump operator is the most important job on the fireground. You must ensure your mates never lose water supply, especially if they are inside the building, as their lives could depend on it. However, I was champing at the bit to get stuck in and didn't need asking twice.

I told Rick where the hoses were going and who was on the branches and left him to it. I went around to the back of the house and found John, stood slightly back from the house watching what was going on and, now and then, giving the odd instruction

"Station, I'm free now. Rick has taken over the pump, what can I do?"

"Oh, he has, has he?" came the slightly cynical reply, John knowing exactly what Rick's motive was.

"Take that branch and keep a jet on the roof to the left" John instructed me. I picked up the hose and branch and just about got the jet of water onto the roof. The pressure wasn't great because we were only being fed by LPP's and not by the major pump on a fire engine, but it was just about good enough.

After about half an hour of holding the branch, I was getting a bit tired. Holding a hose with a large volume of water going through it is heavy work.

An officer, whom I had not seen before, came over to me and introduced himself as DO Gray. He suggested, as we still didn't have any ladders at the scene, that I try to break a window in one of the gables with the water from the hose in order to get the jet up and under the roof to where the fire was spreading. I directed the water onto the glass, but it had no effect on the panes at all.

"Knock off the jet a minute" he said as he was wandering

around me looking at the ground. He stopped and bent down and picked up a stone 2-3 inches across. "Let's try something a bit different," he said, and, with that, he threw the stone at the window. I have to say he was a rotten shot and it fell well short. With a slightly embarrassed look on his face, he turned to me. "Come on then help me out here". With that I dropped the branch and picked up the nearest rocks and let fly. The first couple followed a similar trajectory to the DO's and were rubbish, but they were just sighters. We had the range and let a barrage go. They all hit the window with resounding clangs and cracks. However, that's all it was, the sound of a crack, the panes all remained intact. "Give up, that's not working" he said dejectedly.

As we turned around, I saw some more firemen walking across the sodden fields with more kit, including a triple extension ladder. When they got to us the DO told me to pitch the ladder to the side of the gable and get my jet to work from the side. This, I did and stopped the fire from spreading to the end gable. The following week, there was a photograph of the fire, published in the Henley Standard, and there in the middle was me climbing down the ladder. I'd had my first picture in the paper.

When the scene was declared safe, we entered the house and put out any hot spots in the eaves. I found myself in the room with the indestructible window. I was intrigued as to why it hadn't broken and, as I went over to it, the DO came into the room. Together, we opened the window, and both uttered in surprise "bloody hell". The reason for our shock was the discovery that the glass was an inch thick. We stood no hope in hell of breaking the solid window with the puny stones.

When I came out of the house, John was rounding us all up. It had been over four hours since we had arrived, and our relief crews were now on the scene enabling us to go home. We started trudging back over the sodden fields, which I was much happier doing rather than going back on that little boat.

As we reached the road, we were met by a fleet of fire engines and special appliances, all crowding down the little lane. Being somewhat isolated at the house I hadn't realised how many fire engines had been called to the fire. Before walking back to our engine by the riverbank, we went to the canteen unit. This was a horsebox size demountable unit which was dropped off a lorry. Then, a dedicated crew set up the boiler and the food for the crews. By then, it was well past lunchtime, and I hadn't realised how hungry I was. I was about to be introduced to a culinary marvel which filled a hole. This was a plastic cup of tea which was so strong that you could stand a spoon up in it. But, best of all, there was a couple of rounds of sandwiches. Never had I tasted their like, but they sure were filling. Never have slices of plastic type bread smeared with a quarter inch thick butter and huge chunks of corned beef smothered with tomato sauce tasted so good.

After that hunger hole had been filled, we headed back to B06W thinking that all was well, and we were on our way home. Oh, how wrong we were!

As we neared the engine, which was now on the lane rather than on the grass where Bill had parked it, we saw a rather irate member of the public with a senior officer. There was a lot of gesticulating and pointing at the grass, and the very deep twin tyre tracks running the length of the lawn. Quietly, we were ushered onto the pump, and we drove away. There was a strange atmosphere in the cab, both Bob and Mick were silent. We looked at each other in the back and shrugged our shoulders and said nothing.

I transpired that the perfectly manicured and tended lawn on which Bob had parked B06W to load the boat was due to be used on the following day for a photographic shoot to launch a new luxury car. This was going to earn the homeowner thousands of pounds in location fees, which now would not happen, due to the deep mud chewed ruts and tracks which our engine had left.

The following weeks involved interviews, investigations, statements and eventually a bollocking for Bob for not following brigade standing orders and for taking a fire engine off hard standing. I believe that the brigade paid a considerable amount of compensation to the owner of the grass.

CHAPTER 9
Back to School

Why, oh why, does the alerter always go off on a Sunday afternoon when I'm just about to tuck into a Sunday roast? Oh well, "can't take a joke, shouldn't have joined!"

Parking and locking the driver's door on the station forecourt, I could see that it was a two-pump shout, as those ahead of me were grabbing the second row of tallies. There was one left for me and, as I climbed onto the pump Mick called through to the back,

"Gillott's, smoke issuing."

My old school, I hadn't been back there for nearly five years, despite its being only a few hundred yards behind our house.

We were halfway there when we heard John, on the first pump, book in attendance, but his tone was always the same and gave nothing away as to what he was seeing. This was not always helpful for the oncoming crews, but I was impressed with the way John approached incidents and I still try to emulate this.

As we turned onto the drive, we could see the back end of Lima poking out from between Mrs Dace's pottery room and Mr King's metal workshop.

We dismounted and, as I turned the corner, John called out,

"Darran feed the hose reel up the stairs to Rick and Nick, they are inside, in BA"

Up the stairs? NO! it was Mrs Nicholson's art room on fire. She was my favourite art teacher, and it was my favourite classroom. Her classroom wasn't a classroom as such. It was a wooden beamed loft room accessible only by an external metal staircase. The room was small with not much headroom, but it had character, and I enjoyed many an art lesson in there. Now it was on fire with lots of kids' artwork going up in flames.

I ran up the stairs, my heavy metal toe capped boots resonating on the metal and started pulling more hose reel up over the railings and feeding it into the room. I couldn't see either Ric or Nick as the smoke was too thick. I leaned a bit too far into the doorway and got a face full of smoke. Coughing, I turned away and, as I looked down, found myself staring straight down my friend Graham's camera lens.

He smiled at me and gave a thumbs up. I smiled and turned away to concentrate on helping Ric and Nick. Eventually, the smoke began to clear, and they came staggering out. Ric, shouting to me through his BA mask said,

"Tell John the fire's out and we've ventilated. But it's gone through the floor to the metalwork shop below".

Because the building was made of the traditional brick and flint construction with no window below the stairs, we couldn't see what was happening below.

I called down to John and repeated what Ric had told me. John acknowledged this and told Bob to go and investigate. Bob came back very shortly afterwards with unwelcome news. He could see burning debris through a side window, but it was all over a gas cylinder.

The job had just taken a nasty development. John ordered the metalwork shop to be broken into and to get a jet onto the cylinder from a distance until we could identify it. If it was acetylene, then we were in big trouble. Most cylinders involved in a fire are relatively harmless and can be cooled by a jet of water and removed fairly quickly. If it was acetylene, because of the way the liquid is stored within the cylinder, it is far more dangerous and can self-generate heat inside. If not cooled for at least 24 hours, it can explode at any time.

We broke in through a back door and fed a hose around the numerous work benches until we could see the cylinder in the distance. Gas cylinders come in many different colours to denote what they are storing inside. Unfortunately, because so much debris had fallen on it from the art room above, we couldn't see what colour it was.

Therefore, until we got confirmation from the school, we had to treat it as worst-case scenario.

Tony and I tied the hose off onto a workbench and asked for "Water on, slowly!"

As the water jet increased in pressure, we adjusted the branch, (the nozzle at the end of the hose), to aim the jet onto the top of the gas cylinder. Once we had the water landing just where we wanted it to be, we backed out of the room and watched the cylinder from outside through a window.

Then we entered a calming down phase. The incident was still ongoing, but the initial rushing about was over. We had just to sit and watch the cylinder until a key holder or teacher arrived to tell us what gas was in the cylinder.

Eventually, Mr King the metalwork teacher turned up and confirmed that the cylinder wasn't acetylene but was propane and was almost empty.

With that, we went in and, after turning the water off, Tony and I carefully rolled the cylinder outside away from the now rather dodgy ceiling above it.

Nowadays, we would use a remote thermal imaging camera to view the cylinder for any hotspots and to get a temperature reading.

But then we just took our gloves off and felt the body of the cylinder, which some of us old hands, so to speak, still do today! Fortunately for us, it was stone cold.

As we started making up our equipment, the police arrived and very quickly established that there had been a forced entry and that the fire was almost certainly arson.

It was now well past 8pm. I was tired, dirty and hungry and my nice roast dinner from earlier on would now be in the bin.

We left the police to their investigations and headed back to the station. As we pulled up on the roadway alongside the station Graham jumped out of the cab to open the station and the appliance bay doors so that we could reverse in. While we were waiting for this, we also climbed out of the cab and stood around chatting on the drill yard.

"STOP HIM, STOP HIM, SOMEBODY STOP HIM!" came the cry from behind us up the street. It was Dave the landlord of the Rowbarge, our local pub. I turned around to see a lad of maybe 16 or 17 running down the hill towards us, with Dave in hot pursuit.

Without knowing why, he had to be stopped, and without really thinking, I stuck out my right arm to stop him. He collided with me at full speed and, as he did so, I grabbed his arm. Instantly, I felt a terrible ripping sensation in my shoulder and lost a lot of my strength in my arm. I though "you bugger, I'm not letting go now!"

We spun round and landed up in the thorn bushes at the bottom of the yard. Some of the crew ran over and pulled him off me and helped me up. My arm was hanging loose by my side. At the time, I didn't realise that my shoulder had been dislocated. I grabbed my shoulder with my left hand and pulled it tight.

There was a click and some of the pain went away. Inadvertently, I had put my shoulder back in place. It was still very sore, but it was moving again.

Dave caught up and said that the lad had broken a window in the pub and then done a runner.

The lad had finally given up struggling when he realised, he wasn't going to win against five burly firemen. With Dave in tow, the lads frog-marched him down the road to the police station, which was as the bottom of the street. I decided to follow to report my injury. At first, the desk sergeant wasn't happy having 5 filthy firemen in his nice clean front office accompanying a lad on only the minor misdemeanour of breaking a window. However, when they told him that he had injured one of their own, me, his face changed from apathy to enthusiasm as he now had an assault to deal with. Out came a stack of blank statement forms and I was ushered through the office to a witness room. A PC who had been tasked with taking my statement started filling in the headings and then asked me to explain in my own words what had happened.

After an hour or so, with hunger pangs getting worse and not having even been offered a cup of tea, I signed the final page and got out and back up the hill to our station.

"Great, at last, dump my kit and get home. Oh no! John was still there waiting for me to return so that he could fill out the fire service statement as I had been injured on duty!"

I had to repeat it all and, after signing yet more forms, I got to go home. I refused to go to hospital to get my shoulder seen to, playing down the injury just to get home. That was a big mistake. I didn't realise, but another circle in my life had just started. My shoulder had been weakened and would frequently pop out in the future, so much so that a year afterwards I had to have major surgery on it. But at the time, I just said that it was a bit sore, and I left it at that.

When finally, I got home and told mum all that had gone on, playing down even more my injury, I got fed and went to bed.

The outcome of that evening was that a couple of months later I was summoned to the local magistrates court as a witness in the assault and criminal damage case brought against the lad.

Nervously, I was sitting in the waiting area when a well to do man came over to me and asked to see me outside. He was a tall man wearing a tweed suit, a sort of country gent obviously with money. He explained that he was the lad's father. His son was applying to go to Sandhurst and any court case like this would seriously damage, if not stop, his career and he didn't deserve that. He then offered me £200 pounds to drop the charges.

I was taken aback by this proposal.

The lad was underage and should not have been in the pub in the first place. To think that daddy would just buy him out of trouble, no matter what the circumstances, absolutely disgusted me.

However I calmly bit my tongue and pointed out that it was the police and not me who had brought the charges, so there was nothing that I could do.

The case went ahead. I was questioned about my injury and asked if I had made a full recovery, I said that I had. At the time I thought I had but, had I known the lifetime of hassle and pain which I would endure then, my reply would have been different.

His mitigating circumstances, sob story to you and me, about Sandhurst and a minor mistake etc., etc won over the bench. The result was a verdict of not guilty and no compensation. He got away with it and I was left disheartened with my first brush with the judicial system and what appeared to be a class issue.

An issue, shall we say, between a part time blue collar fireman and a potential officer and a gentleman. I will leave it at that.

CHAPTER 10
Decision Made

Three days before Christmas and, once again, I find myself nearly flooding the engine of my Escort desperately trying to coax it into life in the early hours of a Monday morning. Finally, to the smell of an over-rich petrol mixture it splutters into life. Following my now familiar and deserted route, I head for the fire station. As I turn, gingerly, onto West Street, I see B06L disappearing down past the town hall. I park and, as I get out, I get the sinking feeling of not being needed. This is because nobody else was rushing to climb aboard B06W which is sitting quietly to the left of the appliance bay, next to the exhaust-filled space which was recently vacated by the first engine.

Calmly I walk into the muster bay to find Mick and Tony reading the printout.

"It's a roof fire in Russell's Water," Mick reads out. "Watlington and us. Go turn the radio on Darran" he says. With that, I turn around and walk back, through the lingering exhaust fumes from Lima, to the cold fire engine, open the passenger door and climb in to reach the radio. I push the on/off button and turn the volume up in time to catch the tones of Barry Adby, the officer in charge of the Watlington appliance, booking mobile, (to the incident). "Bravo 9 Lima over". If they are only just going mobile, then we will have a good 10-minute wait to hear anything as Russell's Water is right on the edge of our two station grounds and quite a distance to travel. I leave the cab door open and return to the warmth of the muster bay while listening through the plastic flaps in the doorway.

While we are sitting waiting for the first message, Graham with one of the newbies, Pete, walks in. As he sits down, the radio crackles into life.

"HI Bravo 9 Lima Over."

Barry sounds excited and we all go silent, intently listening to the metallic voice breaking the cold morning air.

"Bravo 9 Lima go ahead HI over" comes the calm reply from fire control.

"HI in attendance Russell's Water. Assistance message........."

We all jump up at once and start putting on our fire kit. We don't need to hear the rest of the message as we know we will be the next fire crew into the incident.

When the bells go down, we are already dressed and the engine roars into life and refill the now clear appliance bay with more exhaust fumes. Just as the sounders fall silent, we have pulled out of the station and have the doors closed.

After Mick had booked mobile, fire control reply.

"Bravo 6 Whiskey for your information Bravo 9 Lima has made pumps 8 roof well alight HI over"

As Mick acknowledges this information, in the back, we look at each other with raised eyebrows and grins. It still feels strange that I can smile at someone else's misfortune. Don't get me wrong, I always feel sad for people in distress, but it is what we are trained for and relish the opportunity to put training into practice.

However, I had another thought on my mind. It was four in the morning and, if this job was to go on as I guessed it would, I would be extremely late for work. I knew that we didn't have much on today and I thought I would just wing it when I got back to the station and phoned in.

Sometimes, at night, when approaching an incident, you can see a glow from miles away and the adrenaline starts to flow. In this case, we had turned off the main road, down the single-track lane and had turned past the famous pond (more of that in a while), before we saw the glow and then the flames. The property was an L shaped two storey house, with over two thirds of the roof either on fire or smoking badly.

As we pull up, John is walking over to us. In all the time I worked with him, I never saw him run. It wasn't because he was lazy or unfit. It was because I never saw him panicked or rushed. We are told to set the lightweight portable pump into the pond which we have just passed.

Now, this pond will be familiar to those who are fans of the film Chitty Chitty Bang Bang. It is where Truly Scrumptious (Sally Ann Howes) crashes her car twice and is rescued by Caractacus Potts (Dick van Dyke). In fact, I have fond memories of my father taking me out in our car to see the film being made. The windmill scenes were filmed nearby at Ibstone and we saw some of the filming from afar. I also remember my Auntie Rene taking me, as a little boy, to see the film in Leicester Square, just after it was released.

Now, I was humping a lightweight pump and all the hose to the famous pond. We intended emptying it and subsequently stranding the ducks in their little house in the middle.

We set up the suction and started the pump. The problem was that although the pond was large in area, it wasn't particularly deep and consequently, we had difficulty in keeping the strainer of the suction completely submerged. We had a hundred yards or so to run the hose to the house, so we started this leaving Pete to stand in the freezing pond water with his foot on the suction, keeping it under water until we could get the heavy rubber hose ramps to weigh the strainer down under the water. We reached the fire-ground with the hose and plugged it into Lima and shouted for "WATER ON!". Tony on the LPP obliged.

As we were catching our breath, Mick called to us to get the 13.5 ladder pitched to the end of the roof which was not yet on fire. This was to create a fire break to try and save some of what was left of it. As it needed four men to pitch the ladder, Graham and I looked around for help. We shouted to Tony to help and, as the water was running nicely from the LPP, he left it and came running over. We found a Watlington bloke who was free and, together, we took the ladder from his pump and headed towards the house. We pitched the ladder first time, which is quite usual on the job. I say this because on a drill night we can often take three or four goes to get one accurately pitched but, time and time again, on an actual job, the adrenalin kicks in and the most difficult ladder pitch will go in sweetly first time.

Tony grabbed a fireman's axe and, with Graham footing the ladder and me on a hose reel spraying a jet onto the roof to protect Tony on the ladder, he got to work breaking the roof tiles and making a couple of feet width firebreak stop the spread.

In the distance, we could begin to hear more two tones of the make-up pumps approaching from Wallingford, Goring, Thame and Didcot. As they arrived, one by one, they were greeted by John and, following a brief exchange with each officer in charge, the crew disgorged from the engines and got to work on their designated task.

As the extra crews arrived, we are relieved from our task and can take a quick break at the truck. Now, we've been on the scene for about an hour and a half and, and as the village awakes, numerous trays of tea and coffee begin to appear for us from the neighbours. Tony, Graham and I are sitting on the appliance steps enjoying a nice hot cuppa on this bitterly cold winter's morning when Graham asks, "where's Pete?" There's a pause, then all three of us simultaneously reply "SHIT!" When Mick told us to pitch the ladder to make the fire break, we had all forgotten that we were going to take back some hose ramps to keep the suction submerged in the pond. We all have grins on our faces that broke into laughter as we realise that Pete must still be standing in the freezing water keeping the suction submerged. With that, Tony and I get a couple of hose ramps from the roof of the appliance and Graham grabs another mug of tea and we head back to the pond.

As we rounded a bush, we see this pathetic sight of a fireman, with a couple of mallards for company in the middle of the pond, shivering like mad.

"Where the f*****g hell have you lot been? I can't feel my bloody feet now". As we wade in and place the ramps over the strainer and release Pete to hug his mug of tea we try to explain, through stifled giggles, what had happened. Somehow, I don't think that he believed us.

Now warmed up again, we head back to the house for our next tasks. The roof was now only smouldering, but we could see that almost all of it had been damaged. John told us that, as a group, the main task now was to salvage as much of the homeowner's property as was possible. Their double garage had escaped the fire and we were to put as much of the furniture and belongings as possible into it.

John had asked me to concentrate on the wardrobes, so I headed upstairs to the main bedroom. As I enter the room, I meet a couple of crew from Goring who are pulling down the ceiling to find any hot spots. I tell them what I am doing, and they clear the way to the built-in wardrobe in the far corner of the room. It seemed strange, walking into someone's bedroom and, in effect, rummaging through their personal effects. The fire had spread along the ceiling and had found its way into the top of the wardrobe which I was emptying.

Carefully, I started lifting their clothes from the rail and putting them into the bin bags which I had brought up. Then, I reached up onto the top shelf and felt some boxes tucked to the side. As I pulled them down, a lump came to my throat. The burnt and wet boxes were wrapped up in children's Christmas paper with labels and bows on them. They were presents for the children of the house which had been hidden away from prying eyes and were now damaged or destroyed. Then, it hit me that their Christmas was ruined as well as their home. Carefully, I carried the presents down the sodden carpeted stairs and out into the ever-increasing daylight. A small crowd of spectators was now gathering, and I felt some anger toward them as they gawped at this family's misfortune. I entered the garage and was hit by the combined smoky and musty smell of the furniture which was being stacked up. I put the presents on a pine dresser at the back of the garage and returned to the house to continue my task.

After being on the scene for nearly five hours, John calls together the Henley crews. He tells us that the relief crews are on their way and that we were to make up as much of our equipment as was possible so that, when they arrived, we could get away quickly. Graham, Tony, Pete and I head back to the pond to make up the LPP and the hose.

"Pete can you get the hose ramps out the water" Tony asked with a grin in his voice. I won't repeat Pete's answer here, but I took pity on him and paddled in myself to retrieve them.

After everything was put back on 6 Whiskey and we were standing around waiting for the reliefs, I began to feel really cold. After all the energetic work of the night and morning, I was very sweaty and damp and, with the cold of the morning, I began to shiver. I decided to climb into the rear cab out of the icy morning breeze.

Finally, our reliefs arrived and we headed home, leaving the family to salvage what was left of their festive period.

When we got back to station it was past 0930 and I knew that Ray, my boss, would be wondering where I was. I phoned work and Paula, his wife and business partner, answered the phone. I explained where I had been all night and asked whether it be ok if I didn't come in today? I explained that I was shattered and knew that, as it was so close to Christmas, we didn't have much on today. Nothing that couldn't wait until tomorrow anyway. She agreed to it but, by the tone of her voice, I could tell that she wasn't completely happy with this request.

As most of the rest of the crew had work which they had to

get to, and whilst I now had the day off, I offered to come back to the station in the afternoon to finish cleaning the kit. John accepted this offer and we all left for our homes to freshen up and grab some breakfast.

After a lovely hot steamy shower and a good old-fashioned fry up, I felt much better than I had done as the damp cold smoky person who had slumped through the front door an hour earlier.

I still had the image of the ruined presents in my mind as I drove to work the following day. Nearly thirty years on, it is still as clear today as it was then.

The following day, I hadn't been in the office for more than twenty minutes when Ray called me into the front. Basically, he and Paula then had a go at me for not coming into work on the previous day. They were very calm and decent about it, but they made it clear that they were not happy and felt that my part time work was interfering with my main job.

But this had been the first time that I had not come into work due to a fire and they had agreed to it!

I was really unhappy at this accusation. Had they not agreed then I would have come in. Also, I was angry at their response because had they seen what I had seen, including the looks of despair and shock on the family's faces at the devastation of their lives in front of them then they would not feel so high and mighty in telling me off.

I bit my lip and apologised through gritted teeth and sloped off again to the sanctuary of my darkroom. Sitting in the dark, yet again, in the place where it seemed that I had made all my serious career decisions, I made the decision to leave the photographic world and to become a fulltime fireman.

To start this, I would find a job, any job, in Henley, which would let me respond to the fire calls 24/7 as a retained fireman until I could get in full time in the fire service.

At the end of the day, I wished everyone a merry Christmas and headed home for the festive period and to a copy of the Henley Standard, not just for the news of the fire, but for the situations vacant on the back page.

When I got home, I told mum of my career decision. To her credit, she didn't point out that I would be giving up a really good job for something which would enable me to do a part time job. No, she just asked me if I was sure that was what I wanted to do and if I would be happy. "Yes" was my answer to both.

It being Christmas, there was no Henley Standard that week so I would have to wait until the New Year to start my job search.

Christmas day, and I awake to a chilly morning with a thick diamond frost over the garden. Not a white Christmas, but close enough. It's going to be a quiet Christmas with just me and mum on the day and my brothers visiting over the festive period. Nowadays, as a husband and parent, I always cook the Christmas dinner and insist that my wife, Sheila, does absolutely nothing.

However, when I was at home, mum always did the cooking, and she made a mean Christmas dinner with all the trimmings. One o'clock arrives and the table looks superb. We sit down and start to dish out the turkey, crisp roast potatoes, parsnips, sprouts, carrots, peas, pigs in blankets, cranberry sauce, stuffing and lovely rich gravy. My mouth is watering with eager anticipation.

As we reach out to pull our crackers, instead of the familiar crack of the banger inside, the dinner is wrecked with the

BEEP BEEEEEEEP BEEP BEEEEEEEEP BEEP BEEEEEEEEEEEEP of my alerter going off.

"Shit" I shout out at the top of my voice.

"No, surely you don't have to go now Darran?", mum says despairingly.

"We're are on minimum crewing, I've got to go."

Minimum crewing meant that we had agreed that those with children could have Christmas Day off and the singletons, or those that really didn't care, would provide cover. It meant that we only had the bare minimum of eight on call, four for each machine so I had no choice. If I didn't go then one machine wouldn't turn out.

As I drive down to the station on the almost empty festive roads, I am cursing my luck and listening to my rumbling stomach. I pull onto the forecourt and jog into the station. I see John getting dressed and looking around to see who was coming in. He must be careful not to take the first four through the door because if he takes the wrong people then there may not be the right people, drivers etc. to crew the second pump. John tells me that it is a chimney fire and to get my kit on. I don't argue or show any sadness, which would not be professional. It is only a one pump shout and I would have been more than happy to have stood down to crew B06L and then go back home to my dinner which, as I speak, is probably getting wrapped in foil and put back into the oven.

Four of us climb on board and we head off to Harpsden Road. Fortunately, it is in town, so we are in attendance in a couple of minutes. As we pull up, an elderly lady comes to the front door of the mid terraced Victorian house.

"I'm so sorry to call you out but it started roaring and I didn't know what to do" she says in a very frail tone.

"Not to worry, that's what we're here for. Now, let's have a look and see what's happening" John reassures her in his usual calming tone.

As we had pulled up, we saw an unusual amount of smoke coming from the chimney stack, so we knew we had a job on. Without being told, we start getting the canvas runner sheets from the engine and lay them in a path from the front door, through the hall and into the tiny front room. Now very carefully, we move as much furniture as possible away from the hearth and then lay a sandwich of sheets over the floor and fireplace. Firstly, we lay a clear plastic sheet to stop any further water damage to the carpet. Then, we lay a leather blanket on top to stop any hot coals which may fall out of the grate from burning the carpet. Finally, a canvas sheet was laid to cover all around the fireplace that we can't move.

This may seem a lot of bother for a small fire, but John had a mantra which he always followed and, to this day as an officer in charge, I try to keep.

John would tell us.

"I have an image of a neighbour coming into the house the day after the fire and saying, I see you've had the firemen in, because of the mess. I would rather they come in and not even realise we had been there." That is why we took so much care to protect as much as possible. In fact, there were sometimes when we left the house tidier and cleaner than it was when we arrived.

Once we were ready, I put together the chimney rods and the old stirrup pump with a hose and metal rose connection on it. With a bucket of water next to us, I slowly pushed the rose, one rod at a time, up the chimney, working the rods from side to side while Rick slowly pulled the handle of the stirrup pump up and down. Strange that, Rick was on the less energetic job again.

As a child, in my Ladybird book on the fire service I had seen a picture of a stirrup pump in use. It was a very old piece of kit. Indeed, the one we were using was brass and stamped G VI R as in made during the reign of King George 6th. However, as I've said before, it doesn't matter how old something is, if it isn't broken and it does the job it is designed for don't fiddle with it.

We got on with the job of rodding and shovelling out all the wet soot which fell into the hot grate.

When we had finished, the lady insisted that we have a cup of tea with her before we left. She brought out a tray with old and chipped floral-patterned cups. On the tray was a matching milk jug, complete with more chips, together with a sugar bowl and a plate of digestive biscuits. I began to realise that this lovely old lady was lonely on Christmas, and that possibly we were the only company she would have that day. We tidied up and then sat down with her for a bit of a chat and a drink of a very strong and lukewarm cup of tea.

Eventually, we made our farewells and left the lady to her house which was now cold as she wouldn't be allowed to re-light the fire until the chimney was swept.

When we got back to station, I went to get the chimney gear off the pump to clean it but, to my surprise, as I did so, John said to leave it for another day.

Gratefully we accepted the order and hung up our kit, signed for the double pay to which we were due and headed home. As I walked through the door, mum got out of her chair and headed into the kitchen to fetch my now somewhat dry Christmas dinner. To my surprise, mum brought through two plates from the kitchen. She hadn't wanted to eat her dinner without me.

As I went to put my knife into the food again, I paused, peace and quiet, so I tucked in.

CHAPTER 11
Full Cover

1987 arrived and I finally got to read the situations vacant in the Henley Standard. The largest restaurant in town, The Catherine Wheel, was advertising for a barman. Not exactly the career move I was looking for, but it would do as a stop gap. One phone call later I found myself in the manager's office trying to explain what a retained fireman was and what I was expecting him to agree to, such as letting me leave work as soon as the alerter operated. After a bit more explaining about how long I could be away for he offered me the job!

The following day I gave my notice in to Ray, much to his surprise. A week later Steve, the other employee in the company, came home with me and after emptying the car at home he drove off in what was my company car.

I had gone from a company professional photographer on £8000 per annum and a company car to a barman on minimum wage with a moped as my mode of transport. Some would say, including my mum, it was not the smartest career move, but I was happy I was full cover.

The bar work was not exactly taxing and whenever the alerter went off it was a bit of a release, and I was usually the first or second into the station meaning I always caught the first pump. My bar uniform included a white shirt, which as you can imagine does not go well with a sooty fire tunic. To solve this, I wore a white tee shirt underneath with a clip-on tie. As I ran up the marketplace, I would be pulling off my tie and after unbuttoning the top two buttons I would pull the shirt off over my head which usually worked well until one crowded Saturday afternoon. It was lunchtime and the town was rather busy as was the bar when my alerter went off. I apologised to my fellow bar person as I always did whilst running out the hotel. I turned onto Hart Street and after crossing Bell Street started up the pathway along Falaise Square dodging the numerous pedestrians. To help clear my route I usually left my alerter sounding to draw attention to myself and hopefully have people move out of my way. I wish that this day I had silenced the alerter. As I was running up the pavement, I had unclipped my tie and unbuttoned the top of my shirt and began to pull it over my head. As the shirt came clear of my face, I was met with the bright red pillar box, that I usually swerved around, literally inches from my face. I had no time to swerve or stop and hit it square on at speed and fell backwards like a cartoon character flat onto the pavement. Slightly dazed to say the least I opened my eyes to see the classic circle of faces staring down at me. I struggled to my feet in front of the gathering crowd. I decided that to stay and try to explain what I was doing was not a good idea, so I continued up the road to the station at a slightly slower pace with a somewhat smarting face. When I got into the appliance bay, I was met by John coming out of the muster bay.

"Darran what the hell has happened to you?" he said trying to hide a giggle.

I looked down at my white, now blood red tee shirt. In my rush to get to the station and the adrenalin in my body I hadn't realised that my nose had taken such a whack that my face and clothing was now covered in blood.

Working behind the bar you do meet some interesting characters. One was a lovely old guy called Peter. He was about

158

60ish, and I never really knew if he worked or was retired but he was in most nights. He always drank Guinness but only from a bottle not draught, and it had to be from Dublin. If the bottle came from the Park Royal brewery in London, he wouldn't touch it. It used to take us ages checking the bottles and reading the labels when he came in, so much so that in the end we kept a few separate on a different shelf for him. Every now and then we would test him by slipping in a Park Royal bottle. He would take one sip smile and hand it back. He could really taste the difference. The other skill I learnt at the Catherine Wheel was how to make a floater coffee. Some skills you never forget!

Valentines' night 1987 is a night I will always remember. As you would expect it was one of the busiest nights for the restaurant and bar. Sure enough, we were packed and when the alerter sounded at around 9 o'clock I was not very popular with my fellow bar staff as I ran out. Even worse was the fact that the manager was in the main entrance way with two alerters approaching him. Little did I know that Rick was in the restaurant with his wife. As we ran together up the road to the station, we heard a plethora of sirens heading out of town towards Shiplake.

"I bet that's where we are going" Rick said.

Sure enough as we got to the station the printer had stopped spewing out its message.

B06L B06W

RTA PERSONS TRAPPED

2 CARS

A4155 OUTSIDE SHIPLAKE COLLEGE

SHIPLAKE

As we head out of town following the sirens, I started thinking about the section of road we were going to. It was a renowned bad stretch which had experienced many crashes. When I was at Shiplake Primary School I was nearly knocked off my Chopper bicycle going to school one day as I crossed it.

As we came up the hill towards Memorial Avenue, we saw the myriad of blue lights that belonged to the sirens we had followed. We pulled up across the road which was completely blocked by a Jaguar and another car. As John climbed out of the cab and walked up to the paramedics to get a briefing, we could see that a lot of equipment would be needed and without being told we slipped into automatic mode and started unloading the cutting gear and more lighting.

There were two males trapped in the Jaguar one in the front and one in the back. As the vehicle was a two door, we would have to make a third door in the side to get the man in the back out. As Nick and I climbed into the car the man said that he could move so we decided to slide the passenger seat forward and with a little encouragement he gingerly slid out onto a stretcher.

"Well, that was easier than we thought," said Nick.

The driver, however, was a different matter. The steering

column had been pushed right onto his chest. Remember this was before the time of driver's airbags, and we would need to pull it off before we could release him. At the scene was something new to us, a doctor. Nowadays we have HEMS fast response doctors etc but in 1987 this was new. Doctor Pimm drove a large Volvo estate with a green flashing light on the roof and we came across him at car crashes etc. He was a person we, as a crew, respected. Often when we ask the ambulance staff how long do we have, to do the extrication? Their estimates were frequently off the mark. Dr Pimm on the other hand was always calm and would give us a workable time. This was before the concept of the golden hour. However, if he said that the casualty should be out now, he meant it.

This evening he had a medical student with him. This man was pushing his way into the car through the driver's door and trying to make the casualty take breaths from the Entonox, a pain killing gas. Every time he put the mask on the man's face, he pushed it away. The student would then put it back on again, only for it to be pushed away again. To be honest this student was a pain and was getting in our way. Dr Pimm saw this and gently put his hand on the student's hand and pulling it and him away said.

"He is in pain but when he wants it, he will ask for it"

Respect, I thought, as he cleared the way for us to do our work.

As the others positioned the hydraulic jaws on the bonnet of the car and attached chains to the bumper towing hook one end and wrapped the chains around the steering column at the other, Nick and I positioned ourselves behind and to the side of the driver to slide him onto a spine board when the wheel was clear.

As the jaws were squeezed together the chains tightened and the steering wheel slowly lifted off the driver's lap. As he became clear, Nick and I slowly turned him around. As we did, he became abusive and started thrashing out with his arms and caught both of us across our faces. Nick quickly put his arm out and calmed him down. Finally, we got him onto the board and slid him out through the passenger door to the ambulance crew waiting on the road.

Once we got out of the car Nick and I sat on the verge looking at the scene. The ambulance was parked up for some considerable time before it finally moved off with its blues still going, heading for the Royal Berkshire Hospital in Reading.

"He was a bit lively", John said to us as he walked back to the engine.

"You can say that again, do you think he'll be ok?" I asked as we gathered up out kit and followed him.

"I should think so" John replied

That Friday I got a shock when I bought a copy of the Henley Standard. The driver was DOA. He had died on the way to the hospital. After all he had been through and all the thrashing about, he had died. This was the first time, but not the last, that I would get shocking news from the papers about an incident I had been at.

For the first time I began to wonder why I was doing this. We bust a gut to get the driver out alive and it looked like we had succeeded, only to find out he died. I was somewhat demoralised as I read the front-page story.

As I sit in the station mess room of Hay Mills fire station writing this chapter at 2 in the morning, almost 30 years later, it is still as clear in my memory as if it were yesterday.

A few weeks ago, I was reminded exactly why I do this job and why I love doing this job. I now work as a full time Crew

Commander for West Midlands Fire Service, and it was on a dark November evening as I was driving home from a swift water rescue course. I had left the motorway and was on the back roads heading towards Banbury near to the Kineton military base. As I was approaching the hill that led to a railway bridge, I could see a line of brake lights of stationary vehicles up the hill. On the other side of the road were cars turning around and coming back towards me flashing their headlights. I began to get that sinking feeling.

As I reached the end of the line a man was walking down the road leaning into each driver's window and then the vehicle began to turn around. Finally, he reached mine.

"You better turn around mate there's been an accident and you won't get through."

"Is anyone hurt?" I asked.

"Yes, there's someone trapped in a car"

"Are the emergency services there yet?"

"No mate" he replied.

"I'm a medic trained fireman you better let me through"

"Too right mate off you go."

As I drove past the queue being careful not to be hit by any of the U-turning drivers, I began to plan my actions in my mind unsure as to what I might meet as I crested the railway bridge.

As I parked up, I could see a small hatchback with two soldiers from the base at the driver's door. On the other side of the road was an Audi estate with some people sitting in the open back.

I opened my boot and took out my Hi-Viz jacket and fire extinguisher. I cursed that I did not have my community first responder First aid kit with me. I walked up to the soldiers and introduced myself. As one of them stood up I saw for the first time a girl sat in the driver's seat. She was clearly severely injured. The soldier briefed me on what was happening. The girl has a clear fracture of her lower right leg, and possibly a lot more. She had been trying to get out of the car, but they had kept her in her seat and as still as possible.

I told them to stay there, and I then had a quick walk around the scene and confirmed that although there were other people injured the girl was by far the worst.

I returned to the car and tried to open the rear doors but they were all jammed from the impact which had been very hard as the air bags had operated and all the doors were distorted.

As I came around the other side of the vehicle the first paramedic arrived in a car. I walked over to him and after introducing myself and briefing him as to what I had found I returned to the car, and he started to assess the girl.

Because of where we were I knew what fire crews would be attending and who the OIC of the first one should be, my good friend Tony Thornton. Sure enough a couple of minutes later two fire engines appeared around the bend and Tony climbed out of one of them.

"Hello Tony" I called out.

"Daz what the bloody hell are you doing here?" came the surprised reply.

"I'll tell you later mate. The priority is that car over there. One female trapped and seriously injured. All the doors are jammed. If you can get one open I'll climb in the back and do the casualty care, you'll need all your crew outside."

"OK mate" he replied and detailed one of his crew to pop a rear door.

With a casualty in a car one of the first things that you need to do is to get someone in the back seat to hold their head still in case of any neck or spinal injuries. Doing this you immediately lose one of your crew as once in they also cannot move, so to utilise me made sense and Tony knew I knew what I was doing. With the glass broken and the door opened from the inside I carefully climbed in over the broken glass taking care not to rock the car.

"Hello I'm Darran. I'm just going to put my hands on your neck to keep your head still".

With this I slid my arms either side of the head rest and made myself as comfortable as possible as I knew I would be in that position for some time.

I looked at the paramedic who was kneeling in the doorway.

"What's her name?"

"Jennie" he replied

"Hi Jennie, as I said my name's Darran, and I'm going to stay with you here until we have got you out, ok?"

With that she tried to nod her head in understanding.

"Right firstly don't nod or shake your head just say yes or no. Now there is going to be a lot of banging and popping and noise all around us. Don't worry about it, I'll tell you when it's going to happen, so you don't jump OK?"

I heard a faint "ok".

A second paramedic climbed into the front passenger seat.

"Hello mate, are you ok, I'll take over from here if you want?"

"No it's ok. I'm comfortable here and I think you will be needed there" I replied. There was no way I was letting go now.

Outside I could see and hear the cacophony of activity. The placing of blocks underneath the car to stabilise it and prevent any unnecessary rocking of the car to help protect any possible spinal injuries Jennie may have.

The crews started a "peel and reveal" which is the removal of all plastics inside the car to show any airbag charges so that they don't cut through them when removing the roof. I had already had a quick look around the car and told them, from the inside, where all the SRS (secondary restraint systems) or airbags were so they could find them quicker.

There were shouts of "BREAKING GLASS" as the windows were pinged and broken in a controlled manner. Tony stuck his head in the car.

"All ok Daz?"

"Yeah, are you going for a full roof removal?"

"Yes mate, best option"

"OK fine. Can someone pass me in some goggles and a sheet to cover us mate"

"Yep no probs" and with that he was gone.

I turned to the paramedic in the car doorway.

"We're going for a full roof removal, OK"

"Yes fine now but look for a rapidex please. If her BP drops again, we will need her out asap".

Damn, that was not good news. On RTC's we have a plan A which is not time critical, i.e. the roof removal, and a plan B or rapidex (rapid extrication) when the casualties condition goes downhill fast and must be out of the car rapidly.

I looked around and saw the Crew Commander of the second pump outside my rear window and called him over.

"We need a rapidex ready can you sort it" I said quietly so as not to alarm Jennie.

He looked at me and then at the driver's side of the car.

"Doors off and B post out?"

"Yeah, sounds good to me" I whispered.

"Ok leave it with me" he said quietly.

I looked at the paramedic and he had heard what we had said and quietly nodded at me in understanding.

If Jennie's condition dropped again then the side post and driver's door would be off, and we could spin her out. Not ideal in any way but needs must.

With that the crews started cutting the roof off. There was lots of banging and popping of metal, but the cries of "CUTTING" told me when it was going to happen, and I could then warn Jennie. The windscreen was cut through and the final post cut.

"Ready, brace, lift" was the call and six fire-fighters lifted the roof up and carried it forward and off the car.

Finally, the paramedics had space to work.

With the second paramedic taking temporary control of Jennies head from the front, so I could remove the headrest to get access to her back, I took back control of her head and together we placed a cervical collar around her neck.

We could now see clearly her legs. Her right had open fractures and her left leg and ankle were not in natural alignment. The paramedic said he had given her two doses of morphine, but it wasn't having any effect on the pain.

I explained to Jennie that we were going to slide a long board down behind her back and once we had the seat lowered, we would lower her back before sliding her up the board.

The first paramedic and I both said at almost the same time.

"Jennie there is no other way of doing this. It is going to hurt."

She whispered an acknowledgement.

It doesn't matter who you are, what your job is or your rank, if you are holding the casualties head you are in charge of any movement. That was me.

The board was positioned and slowly slid down the seat. Then whilst it was held vertically the seat was wound back. All this time I was holding Jennies head and neck. On my command the board was lowered back as far as possible. This made her cry out in pain, and I knew it was about to get a lot worse.

As soon as the paramedics and some fire crew had hold of Jennie I said,

"Is anyone not ready?"

All said "Ready"

"I will say ready, brace, slide and we will move her 6 inches up the board, ok?"

"Ok" was the chorus.

"Ready.......Brace........Slide"

A blood curdling "AAAHHHHHH!" came from Jennie, with several expletives.

"I'm so sorry" she said

"Don't you worry girl, you shout out and swear as much as you want" I told her.

"It's my leg, my leg".

We could all see her left leg was in a bad way, but we had no choice we had to move her again.

"Is anyone not ready?" I asked again.

"Ready"

"Ok another 6 inches, ready........brace........slide"

This was met with another scream from Jennie.

We needed one more move.

"Jennie this is the last one I promise you" I quietly said to her.

"Ok last 6 inches, anybody not ready?"

"Ready"

"Ok last 6 inches ready.......brace........slide"

That was it, she could relax a bit.

The bright orange foam blocks were fastened either side of her head and I could then after over an hour, let go of her neck.

She was then lifted onto a stretcher and taken away to the awaiting ambulance.

With everything calming down I had a chance to chat with the paramedic who was now packing up his kit.

He suspected multiple leg fractures, and called them "life changing injuries," he wasn't wrong.

As the crews were making up the kit, I had a chance to have a chat with Tony and finally explain how I came to be there. The officer in charge of the incident came over and thanked me for my help. With that the police let me drive through the cordons and finally head home.

A couple of weeks later I found out from Tony that he had found out some information on Jennie. She had been kept in a coma for days while they operated on 20 or so fractures.

4 weeks later I saw a picture of Jennie on Facebook. She was sat on the side of a hospital bed in a full body harness with neck brace, left leg in plaster and right leg in a cage. It was the first time since the accident that she had sat up with a lot of assistance.

Now I was to break a golden rule of mine, I contacted a casualty. I arranged to visit her in hospital and meet her mother as well.

I found myself standing outside the ward entrance, strangely nervous. When I entered the ward, I was met by a beaming smile. We spent about an hour talking about the accident. With her mother's approval I was able to tell her what had happened after I got in the car.

The shock for Jennie was that we, as in the crews, would not have been surprised if we had heard that she hadn't survived.

Her mother agreed that in the first few days no one was sure she would survive.

I came away from the visit with an overwhelming feeling of admiration and respect for her courage and positivity for the situation she found herself in and the long road to recovery.

What shocked me was when she told me that she had 3 fractures of the spine one of which was an unstable C spine injury, the exact place in her neck that I had been holding for over an hour. The surgeons had told her that if that had moved it would probably have killed her. That took my breath away. Also, if we had had to carry out a rapidex we probably would have caused more serious injuries or probably even worse.

The fire service has often been criticised for being over cautious when handling casualties in cars. Well, this was the second incident that I have been involved in where being over cautious has saved a life.

This did make me feel good about myself and the reason why I still love this worthwhile career.

CHAPTER 12

Dangerous

Wednesday 18th March 1987. The realisation of what I was doing was about to hit home. 1800 and we all begin to wander through the door to the station. As usual, I sign in the pay book and then sit on the muster bay platform and generally chewed the fat with the rest of the lads as they meandered in. The usual volume of chat gently rises as more arrive and then, looking at the clock 18:29 we all make our way through the hanging plastic flaps to the appliance bay and line up waiting for the first parade. Strangely Mick Goddard wasn't standing out the front to call us to attention as John would walk through. None of us had noticed that Mick hadn't been in the muster bay either. 1830 arrived and went and still no sign of Mick or John.

Finally, they both appeared through the flaps and Mick called us to attention.

Something was wrong. The expressions on John and Mick's faces I had not seen before. The atmosphere suddenly cooled as John cleared his throat.

"Stand at ease" he quietly he said, almost apologetically.

We all shuffled our feet glancing at each other and then looking back at John.

"Lads I've got some very sad news to pass on. I've just had a phone call from HQ informing us that Sub Officer John Wixey the officer in charge of Charlbury retained has been killed at a fire this afternoon and several of his and Chipping Norton's crews have been seriously injured."

The whole line of us seemed to take a simultaneous intake

of breath followed by numerous hushed expletives. John continued.

"We don't know the exact conditions of the crews but what we do know was that there was a fire in a quarry and whilst dealing with it there was an explosion. That's all we know at the moment."

John's voice remained quiet as he said.

"I suggest we all have a minute silence for Sub Officer Wixey and his family as well as for the other injured firemen and their families."

Without saying anything else we all stood back to attention and bowed our heads.

My mind started to wander. This man had left his family this morning when his alerter had gone off, never to return. I know how that feels. Images of my dad stood in the doorway in his oilskins watching Tom and Jerry with me, then turning and walking out the door flooded back. Now I was doing the same to my mum when my alerter sounded.

Everyone in the appliance bay was stone silent and I'm sure we were stood for far more than a minute when John again whispered,

"Thank you, gentlemen, dismissed."

With that we all turned to our right and quietly went about our checks.

The rest of the evening was somewhat stilted, and I was grateful when half eight arrived and we all went upstairs for a drink.

The atmosphere up there was still very quiet and when everyone had their first drink, we made a toast to John Wixey and his family.

When I got home that evening my mum could see something was wrong and when I explained to her what we had been told she also went quiet, and I could imagine the same thoughts going through her mind that I had experienced earlier.

That night and the following day the thought of what had happened was still playing on my mind and I was willing for the next shout to come. It didn't for two days.

I was wiping up glasses behind the Catherine Wheel bar when my alerter burst into life. My stomach churned and I nearly threw up as I dropped the cloth and called out to my manager as I ran out the door wondering what I was going to and if I would be coming back. I must have run faster than normal as I was first into the station and ran straight to the printer to see we were going to an alarm actuating. I gave a little sigh of relief that it wasn't anything more serious.

The following drill night John had more news and details of the funeral. It was to be a brigade funeral with full honours and representatives from all stations were invited. Nick and I immediately volunteered to go without thinking of how to get time off work. When I explained to my boss the circumstances, he couldn't have been more accommodating and gave me the time off without a second thought.

This was the first time that I was to wear my number one uniform. There is a saying that

"You only wear your number ones for weddings, funerals and bollockings."

This I have found to be true with one exception, Remembrance Parade. The evening before I made sure my blue shirt was crisp and smooth. We were issued with two pairs of best trousers and one pair I had not worn so the creases in the front were good, but I used my Air Cadet service experience to make them razor sharp. I also utilised this knowledge to give my shoes a mirror finish.

The day arrived and John, Nick and I all met up at the station for John to drive us up to Oxford. The arrangements were for all attending firemen to meet up at The Slade fire station in Oxford and then to be bussed to the Oxford Crematorium. When we arrived at the station, the drill yard was heaving with firemen milling around and shouting out to friends that they hadn't seen for a while. As we pulled in, we were directed where to park and when we had, we headed into the station. This was the first time I had been into a full time fire station as an equal not a trainee and even though the occasion was sombre I still felt a little like I had way back when I had visited Didcot with my primary school, slightly in awe but fascinated about this same, but different world, I that I was becoming used to.

The Tannoy crackled into life, and we were summoned back out to the yard where a fleet of coaches were now waiting for us. When we arrived at the crematorium there were marshals directing us where to go. The driveway from the road to the chapel was a few hundred yards long and we were to line the route either side from the gate. As more and more firemen were disgorging from the coaches, we were slowly spacing ourselves backwards and forwards and sideways until both sides of the driveway had firemen evenly spaced along the full length.

The only funeral I had ever been to was my dad's and I really did not know what to expect.

It was a brisk sharp spring day. The crematorium was on a hill and there was a slight breeze blowing up the hill which cut through us like a knife. While I was looking around and the hundreds of firemen, and some fire control women, a shout of

"Parade, Parade SHUN!"

came from our left. The cortege had arrived.

We all glanced toward the gates. The undertaker complete with top had and cane was slowly walking down the drive leading the Charlbury fire engine which was glistening in the spring sunlight. The crew must have spent hours and hours polishing the vehicle and they had done themselves and their station proud. What I wasn't expecting was that the blue lights on top and on the grill had been covered with black cloth as a mark of respect.

There were firemen sat in the engine except in the Officer in Charge seat, John Wixey's seat, which was left empty. As it passed me, I could see that the men were covered in bandages, and I found out later that they were the firemen that had been burnt in the explosion. They reminded me of the TV images of the wounded and burnt from the attack in the Falklands on the Sir Galahad and Sir Tristram.

Following the fire engine was a Turntable Ladder appliance. Again, it was glistening in the sunlight but also with the blue lights blacked out. Stood on the back of the vehicle were four firemen in full ceremonial uniform with gleaming brass helmets, two either side of and holding onto the large mechanical ladder. Underneath the ladder was the coffin draped in the union flag. The vehicles slowly passed us followed by the limousines with the family.

We remained at attention until the coffin had been removed from the Turntable Ladder, and the family had gone into the chapel.

"Parade fall......OUT" echoed through the silent grounds and we smartly turned to the right and broke rank. We were ushered down to the courtyard to listen to the service through loudspeakers relaying it outside.

Once the service was over, we were coached back to The Slade for a warming cuppa and sarnie.

This fire service funeral was my first but hasn't been my last. I have travelled to represent the brigade at brigade funerals over the country. Fortunately, they are few and far between but every time I am taken aback by the number of people that travel from far and wide to show solidarity with the family that is the fire service.

Later that year the fire service family was hit with another fatality in an incident that made worldwide news.

Again, it was a Wednesday and we had just finished drill and were up in the social club having a pint and playing darts when someone said

"Bloody Hell! Shut up everyone and someone turn the telly up quick."

What he had seen was the first live coverage of the Kings Cross fire. The pictures of dozens of fire engines crowding the streets and the myriad of blue flashing lights will forever haunt me.

There were already reports of fatalities coming in. As we continued to play darts and cards, we kept the volume up and then the first mention of a fire service fatality was made.

Station Officer Colin Townsley, yet another Officer in Charge, had been caught up in the rapid fire development underground.

In my first full year the enormity of what I was embarking on had well and truly been brought home to me. This was dangerous!

CHAPTER 13

BA

It was now just over a year since I joined up. I had been to many shouts and had learned a lot, but still I felt that I was only playing at it. I was still having to stay outside whilst the lads in breathing apparatus (BA) went in and did the business.

I was also getting tired of working behind the bar at the Catherine Wheel. Not having my evenings free was becoming a bit of a drag. I saw in the paper a job for an assistant in a fine china and gift shop. The shop was a family business run by the Hickman's.

After the usual interview with me trying to explain what was involved by hiring me, I got the job. There was a slight problem though, 3 weeks after I was to start, I would need 3 weeks off to attend my BA course followed by my HGV driver's course.

In the weeks running up to my move to the china shop I had been applying to fire brigades to become a full-time fireman. I had been successful with Royal Berkshire and was invited to a selection day.

I arrived at the Royal Berkshire HQ in my one and only suit but with a bag of tracksuit and trainers as instructed. There were another 10 or so similarly nervous men in the reception all clutching their sports gear.

We were summoned through some double doors and introduced to a room of officers. They explained that we would be facing a series of practical and physical tests during the morning. We were then led to some changing rooms and told to be outside on the drill yard in 5 minutes.

Once we were all outside, we were divided up to rotate through 5 testing areas. The first I went to was hose running. This I completed easily, running out and rolling up a length of hose 5 times in 5 minutes. One of the assessors walked up to me once I had finished and quietly said to me,

"You're already retained, aren't you?"

"Yes Sir" I replied even though he was only a Leading Fireman.

"It shows lad, well done"

That boosted my confidence for the next test. As I was walking up to it a voice said,

"Is that you Darran?"

I turned to see Alan from the boat yard I had worked at during school holidays

"Hiya Alan, what are you doing here?"

"I'm helping run this today. Where are you off to now?"

"Number 2 station," I replied.

"That's mine, come on let's get going".

Alan's station was a Krypton Factor style test. On the floor were lots of tubes, pipes and valves that needed to be assembled in as quick a time as possible. Alan was looking at me and asked

"Why are you grinning like a Cheshire cat?"

The reason for my joy was that I had recognised the equipment on the floor as a Blackhawk Ram assembly, exactly what we carried on our fire engines, and I had been trained to assemble this with my eyes shut in case I had to do it at a car crash at night.

"No reason" I said in a slightly high-pitched voice.

"Hmm, ok. Off you go then" Alan said as he started his stopwatch.

With that I dropped to my knees and started manipulating the metal tubes, collars and hoses with the dexterity and air of confidence that calmed me.

"Finished" I called out as I moved the handle of the pump to show that it was all correctly assembled and working.

"58 seconds Darran! How the hell did you..............?" Alan paused and the puzzled look on his face turned to one of recognition as he twigged,

"You carry these on your engines, don't you?"

"Might do", I replied mischievously.

Alan smiled "Nice one, well done. That's the fastest time today and I doubt it will be beaten".

I moved onto the other tests including wearing a BA set in what's called a rat run. This is a complex of metal cages that you crawl through. Your pathway goes in all directions up, down, through tight holes in space that you can hardly turn in. This would be good practice for my BA course in Oxfordshire in a couple of weeks' time.

With all the tests completed I left with a happy feeling that I had done my best.

A couple of days later a letter dropped onto the door mat franked Royal Berkshire Fire Service. I nervously opened it to read

Congratulations you have successfully passed the selection tests and we would like to invite you for an interview with the training school panel 0900 on Tuesday 10th November at 1015. My joy at this news quickly turned to dismay. I was to begin my BA training on Monday 9th dam!

The following drill night I went to see John in his office and explained my predicament. John was aware that I was applying to other brigades and told me to write a MIS3, the number for a small A5 pad that the brigade had all small memos written on. I was to explain the situation and send it to the training school commandant.

This I did and after John put his endorsement on the bottom, we sent it to training school at Didcot. By the next drill night, I had a reply saying that I could take the time off for the interview, but I was to report to the training school as soon as possibly afterwards.

Over the previous few months Graham and I had started our BA training on station. There was a set 21 hours of subjects we had to cover and pass before we could attend training school.

The set was very similar to a scuba divers set but it was not designed to go underwater. It had a metal back plate which had webbing straps, much like a rucksack, at shoulder and waist height. From the valve on the bottom of the set came a rubber hose around the side and up to a face mask which covered your whole face and under your chin. This had a slightly inflated rubber seal so that when you pulled the mask on it made an airtight seal around your face. This was the reason, and still is, that fire fighters are not allowed to have beards, only moustaches as otherwise air would leak out around the edges because the facial hair would prevent an airtight seal. There is a saying that you can always tell when a fireman is on leave, he is covered in stubble.

The cylinder was light grey with black and white quarters at the neck to denote it was carrying compressed air. It was made of cast metal and although I can't remember the exact weight I can say that after having it on your back for 20 plus minutes it weighed a ton! All the modern BA sets and cylinders now are a mixture of carbon fibres and are so much lighter.

We had to learn all the components of the set and what they did and how they worked. We had to learn how to operate the DSU, Distress Signal Unit which each set carried.

This was a metal tube about 5in long with a rubber cap on both ends. In the small end there was a metal button which had a small key attached to a bright yellow tally. On the Tally the wearer would write in chinagraph pen his name, the BA set number and the pressure of air in the cylinder. When a fireman went into a fire situation, he would pull the tally and key out of the DSU which would then arm it and hand the tally to the BA Entry Control Officer. If whilst in the fire the fireman got into difficulties and needed help, he would push the button and the bottom of the DSU has a speaker and an alarm would sound signalling to all in and out the building that someone was in distress and needed help.

The set also had some rope in a bag attached to it. The rope was divided into two sections. The first 1.25m long and the second 4.75m making 6m in total. When you went into a large or very complicated building you attached your personal line to your partners BA set, or to a guideline, by a carabineer clip and loop on his set. Initially only the 1.25m so you kept in touch with each other in the dark. If you needed to search a room, you could pull a toggle and release the longer 4.75m length so you could spread out the full 6m. This usually resulted in a tangle of spaghetti, and you had to be very careful controlling the line. This would take weeks and months of practice. Nowadays we have self-recoiling lines which are better, but you can still get in one hell of a mess if you loose concentration.

Not only did we have to learn the set, but we also had to learn the BA entry control procedures and how to run the BA board.

The BA board is a method of recording who is in the fire situation, their location or task, the air in their cylinders and using a special calculator their estimated time of whistle and when they should be out of the building. The BA board is a vital piece of equipment and would get impounded if anything untoward happened to a fireman and could become a legal document and be produced in court, so we had to get it 100% right. The person running the board would put on a yellow and black tabard denoting his job and ensuring that he could not be told to do anything else by an officer.

The other vital piece of equipment we had to master was the BA guideline and its tallies.

The idea behind this was simple enough. You had a bag of line on your back and as you entered a large building it was tied off at the doorway and it then paid out as you walked in so you could find your way out, simple? NO!

Each line had two knots at intervals and lines attached so that in the dark you could find the knots and these would indicate to you the way out. One was a long line with a single knot and line and one was a short line with two knots. There were lots of ways to remember which way was which. "Get knotted" was one meaning. To get out you went in the direction of the two knots. My method was "Longest way in, Shortest way out".

Attached to the guide line at the start was either an A or B tally to denote which line was which if two were in use. Along the guide line you could "Branch off" with another line using the tallies with 1,2,3 or 4 holes in. All these lines had special ways to tie them to doors etc, and complicated procedures to search while attached to them which we had to learn before the course.

These lines and procedures are still in use today as they are simple and work!

With all this training already behind us Graham and I reported for duty at the training school at Didcot on Monday 9th November at 0830 as instructed. Almost immediately 2 bombshells hit me. Firstly Sub Officer Hemmings, my nemisis of basic training, was still there. Secondly to add to the pressure of the course and the time off I was going to have to take, I was made course leader!

Station Officer Henry Moors was still in charge but Sub Officer Mick Sadler had been replaced by Sub Officer Dug Ball. Dug was a large man but with a quiet and mild manner, chalk and cheese with Hemmings.

We settled down to classroom work on the first morning. We went over the sets again and their servicing. After tea break we were all allocated a BA set. I, as team leader, was told to get the sets serviced and ensure that we were all to report in full fire kit, with our BA sets to the garage at the bottom of the yard in 15 minutes.

This garage had inside it a rat run similar to the one I had been through in Reading. One by one we crawlled through the entrance in the bottom and followed the tunnel in the dark. We went up, down, sideways curlled up in a ball to contort around corners, all with a BA set on our backs. After 20 or so minutes one by one we began to appear at the exit.

Once we had removed our masks and retrieved our BA tallies from the entry control officer we sat down on the tarmac propped up by our cylinders on our backs waiting for the rest to come out. While we were waiting a comotion began to emminate from the garage. All we could hear was Hemmings shouting at someone to get a move on. Soon we began to hear the low pressure warning whistles starting their low quiet sound which would increase in pitch and volume as the pressure in the cylinders dropped further.

Eventually, all bar one appeared from the garage door with all their whistles going full blast. When they took their masks off they told us what had happened.

One of our number had frozen in the rat run and everyone behind him was bunching up. Hemmings had been shouting at him to get a move on but he was stuck and began to panic and ripped his mask off. Dug Ball finally lifted an escape hatch and pulled the maskless fireman out of the tunnel releasing the rest behind him to scramble on through the maze to the exit before their air ran out.

We then saw the Subs leading our BA set less colleague out from the side door and into the station. We quietly took our sets back into the servicing room and once they had been cleaned and recharged we returned to the class room.

As we entered I noticed that one of our desks had been cleared of all the books and stationary. After a short while Stn O

Moors came in and told us that one of our number had been returned to station. This was met with silent glances at each other. That afternoon over a cuppa we were all disecting what had gone on. We all had sympathy for our colleague but at the same time we all agreed that now was the time to find out that either he wasn't ready or would freak out in a real fire putting his partner at risk as well.

At the end of the day I double checked with Stn O Moors that he knew I would not be in the following morning due to the interview with Royal Berkshire. He was aware and wished me good luck.

Tuesday Morning and I dressed in my one and only suit and packed by uniform for later. I arrived at Royal Berkshire HQ nice and early ready for my interview at 0900.

Bang on time I was led into the interview room and faced with 3 interviewers. To be honest the whole interview is a bit of a blur, and at the end I was thanked for attending and was told that I would get the result in writing in a week or so.

With that I got back into my car and made a B line for Didcot and day 2 of the course. When I arrived I dashed into the toilets and changed into my uniform and joined the group in the classroom.

This day was taken up with practicing the "BA shuffle" and using the guide lines.

The "BA shuffle" is the motions we make whilst moving through a blackened out room. You slide and stamp your feet alternativly as you move forward to check for any unstable ground and holes in the flooring. Whilst doing this you keep one hand on the wall and your free hand you move in a vertical motion from your waist to over your head to again detect any hazards or loose cables as you move forward. When you begin to tire this movement can become shortened or forgotten altogether and thats what the instructors were looking for. During these sessions we had our masks blanked out to simulate a smoke filled room.

Whilst some were wearing the BA sets others were taking turns in running the BA entry control board. This was just as important as fighting the fire and no mistakes would be tolerated.

At tea break everyone was asking how the interview had gone. To which I could only answer that it was a bit of a blur, but I thought it went ok.

The end of the day came around all too quickly and Wednesday we found ourselves back in the classroom first thing. We were told that most of the rest of the week would be spent at the heat and smoke house at Rewley Road Oxford, and after the morning break we were Oxford bound.

Rewley Road was the main station for Oxford housing three fire engines a turntable ladder and two specialist units. It was built in the 1970's and was, and still is, a concrete monstrosity of the 70's style.

At the back of the fire station was a two storey BA training block. On the ground floor were two "crib rooms" where fires of wooden pallets would be started and then the tempreature would be monitored throughout the building as the smoke spread to the upper floor. Linking the floors were concrete tunnels, shafts and stairs which all had to be negotiated in the pitch black.

We were led into the control room to see all the safety measures involved in the building. The pressure pads in the floor to trace your movements. The emergency ventilation fan controls in case anything went wrong and the complex had to be cleared quickly.

The one outstanding memory of my first entry into the building was the completely soot blackened walls and the pungent smell of smoke. This is typical of any BA smoke house but to experience it for the first time was memorable.

The other image that has stuck with me and has been constantly reinforced year upon year is when meeting the dedicated BA instructors. You can always indentify a BAI "Breathing Apparatus Instructor" by the nicotine type stain on their skin, (Because they spend more time than anyone else in the buildings either setting up a fire or clearing one out afterwards when they don't need to wear BA. The atmosphere of smoke and soot seems to permanently stain their skin and it doesn't matter how much washing or scrubbing they do it appears to become permenant.

Introductions over we were told to find a BA set from the servicing area and after carrying out the daily checks to meet, fully kitted, at the control room.

Our first session was not to be hard work. We were to enter the main crib room as a group and sit on the floor with our BA on. Then the crib would be lit and we would be able to see a full room fire develop, which normally would have already occured before we arrived at a house for real.

We all lined up at the BA entry board and the instructors checked our fire kit was correct and booked us into the building.

Now a quick explanation of the differences between our kit then and now 32 years later. Then we had the traditional cork helmet, with a leather chin strap and a boot lace design inside to adjust the fit, leaving our ears open to the heat. Now we have designer wrap around helmet with integral goggles for eye protection at car crashes but, more importantly, a gold tinted full front visor to reflect the heat.

We were issued with a black silk scarf, (I've still got mine), to wrap around your neck to stop hot embers getting inside your collar.

Now we have a fully fitted flash hood balaclava type, which you put on before your BA mask. Once the mask is fitted you pull the hood over and make a complete seal around the mask, which your partner has to check. Then the collar of your tunic is wrapped around your neck so that there is no bare skin at all. This flash hood also covers your ears, This is important, and I'll explain in a bit.

The tunic, at the time, was just a sort of donkey jacket, double buttoned and velcroed. This had no real heat insulation and got very heavy when wet. Now the tunics are half the weight with integral cuffs with thumb loops to stop them riding up. Currently my tunic feels like its filled with bubble wrap to keep the heat out. Unfortunately in doing such a good job at keeping the heat out it also keeps heat in, and it is easy to overheat.

The leggings, at the time, were a yellow plastic rubberised material. No heat protection at all, just waterproof. They went very limp and soft when heated!

Now the leggings are made of the same material as the tunics and incorporate kevlar reinforced knees for comfort.

The boots were rubber with steel toes and soles, again no real heat insulation. Nowadays they are leather which is much better.

The other big issue was the gloves. At the time they were what I would call the blue rubber gardener type with absolutely no heat protection. Nowadays we have so many specialist gloves for RTC's, firefighting, chemicals etc. The firefighting gloves are leather or kevlar with integral cuffs that have to be rolled over the cuffs of the tunics.

As you can see back in the 80's we had a lot of bare skin, ears, neck, wrists etc. which had its advantages. We were told by the old hands that when the tops of your ears started to roll or peel backwards then it's too hot and you should pull back, advice which I still cherish today.

With today's kit there is no bare skin whatsoever so I find it is harder to feel the true temperature, for example your ears peeling back. With that in mind new recruits do not have the experience of when to pull back and have a tendency to go a little too far into the fire or danger area and often I find myself pulling them back.

We all shuffeled into the crib room and sat on the floor resting our air cylinders against the soot covered walls. The BA instructor NOT wearing a BA set, then started to tell us what to expect to see and also, when we as individuals got too hot, to get on our hands and knees and crawl out through the door to our right.

With that he lit the shredded paper under the number of wooden pallets in the metal cage.

The fire starts slowly at first with a lot of smoke and the flames starting to lick at the palletts.

Sat watching the growing light I became concious of my breathing and the Darth Vader style noise from my mask, so I try to keep it slow and low to make the most of the cylinder and stay in to see the full growth of the fire.

As the flames crept up the cage and started to flatten out on the roof I begin to feel the first effects of the radiated heat. The instructor , who has now backed up to the open door is now shouting out what is happening and what to look for.

The smoke had long ago reached the roof and was starting to creep downwards but before it reached us the heat layer was preceeding it, and I could feel it slowly roll down over my head, past my exposed ears and over my shoulders. Then I saw something that still hypnotises me in fires today, the dancing angels.

Smoke is the product of incomplete combustion and it can burn. Dancing angels is the romantic name for this phenomenon. Flames can be seen leaping through and igniting in the smoke and it is this that is called the dancing angels. If you allow this to get behind you in a real fire then you could be in serious trouble. That is why we are trained to cool the smoke in a room first so that it doesn't ignite.

Watching the flowing flames coming toward me made me forget about the heat and my cylinder contents for a minute or so.

All of a sudden I became aware of the immense heat surrounding me and the fact that there was only three of us left in the room!

That's enough, I thought, and rolling over onto all fours I made my way out at the same time slightly burning my knees through my leggings from the super heated concrete floor.

Once outside the cool air wafted around me like standing under a waterfall as the air mixed with my sweat drenched body.

Set servicing, lunch and then a little search drill to finish the day off.

Walking through the front door that evening I was met with,

"Darran you stink of smoke, go and have a shower before tea!"

Thanks Mum.

As the week progressed the exercises got physically harder and more complicated with learning cooling techniques, search patterns, emergency procedures, navigating stairs, carrying casualties, guidelines and working in larger teams of threes and fours.

By the end of Thursday we were gearing up for our final assessment on the Friday. We carried out a dummy run and were then called into the instructor's office one at a time for a debrief.

"FIREMAN GOUGH" came the shout from down the corridor and off I went at a fast walk to the office.

As I entered I was met by one of the BAI's and Sub O Bell, I breathed a sign of relief, it wasn't Hemmings.

The Instructor started,

"Well Mr Gough how do you think it went today?"

Now do I say all was fine and then get pulled up on numerous mistakes, or say not bad but I did this and that wrong and then still get hammered. Even though I thought it went alright I opted for the latter.

"Not bad sir, but I think I should have done more gauge checks"

"Yes, you were close to your whistle when you came out, but apart from that it was ok. Your search pattern was ok but you forgot your proximity search when you found the casualty. There wasn't another near by but still you didn't check"

Fair enough, I thought, and said I would remember that.

With that I left the office somewhat happy with myself. Others returned with varying expressions from happy to miserable, ie "keep it up" to "If you don't improve you will fail the course."

That evening was spent reading and re-reading my notes for the exam in the morning.

Finals day arrived and we were directed straight away into the lecture room and sat down behind the exam papers. Heart pounding and with the " One Hour Begin", I turned the paper over. Multiple choice was always a bit of a relief but with so many answers very similar with only one or two words difference I found myself reading and re-reding the questions. After about half an hour people starting standing up and handing their papers to the BAI as they left, I was only just half way through!

With five minutes to go and only three of us left in the room I did my final re-read and with a sigh of relief handed my paper in.

In the mess room there was the usual, "what did you put for number 6" etc. I stayed out of the inquisition just listening from the corner sipping my sugary tea.

Hemmings waddled in and with a show of importance to the duty watch who had just come in for their morning tea and toast, shouted

"Fireman Gough recruits to be downstairs ready to start full PPE and BA sets in 10 minutes"

Without waiting for a reply he swiftly turned and was nearly hit by the double swinging doors from the mess as the last of the duty crew came through. He stopped abruptly and then with both hands thumped the doors back open making them clang against the door stops.

As we all got up and started moving toward the back door from the mess I caught sight of the duty crew stiflying laughs and making certain hand gestures towards the now vibrating doors in the empty doorway. I grinned and one of them winked back at me. I'm not the only one who thinks he's an idiot, I thought.

Downstairs we gathered by the main entrance to the training complex. Nearby the training school fire engine had the pump running. Doug Ball and a BAI donned their BA and went into the building leaving Hemmings to brief us. Graham and I were the first team to enter, which I liked as the crib wouldn't be up to full temperature yet. We got our brief, donned our sets and after testing the hosereel I led us through the heavy metal door into the black oblivion of the BA house.

We began our left hand search with me keeping my left hand on the wall and the hosereel in my right. Graham was either pulling the extra hosereel in behind us or reaching out to our right with his foot to search the corridor.

As we progressed down the hallway the heat began to rise. We reached a slightly opened door and in the gloom I could see the orange flickering glow of the fire in the crib. I told Graham what I had seen and we began the doorway procedure to enter the room.

Once inside I laid flat on the ground with my face on its side looking under the smoke. In most fires the smoke will not quite reach the floor due to the cool air being entrained through doorways etc. As soon as you get through a door you hit the floor to look under the smoke level to see if you can see any casualties. This time I couldn't see anyone. We entered the room and keeping on out knees due to the heat we began moving around the room. I was training the water from the hosereel into the smoke to keep it cool as we searched.

As I swept my right leg around in a semi circle I struck something soft. I stopped and Graham, not knowing I had stopped bumped into me.

Through the muffled BA face mask I shouted,

"Hold on Graham."

I then moved out towards where my right foot was and feeling along the ground felt the leg and torso of a training dummy.

"Casualty!" I shouted and Graham moved toward me to help move it.

Just then Dug Balls words from yesterday came back to me

"Proximity search"

"Graham can you get him to the door way? I'm just going to do a proxmity search".

"OK" came the muffled reply.

As he pulled the dummy along the floor I laid flat to keep out of the heat and swept the area around the casualty. Yet again my foot hit something soft and as I reached out I found a second casualty, this time a child.

"I've got another one!" I shouted to Graham .

"OK, I'm at the door now can you mamage that one?"

"Yes its a child, are you ok with with that one?" I replied.

"Yeah fine, lets get out of here."

With that we both crawled out of the crib room, shutting the door behind us. Now in a cooler atmosphere I could just about see that Graham was struggling with the twelve stone dummy so I said to him to hold the arms and I would grab the legs with one arm and hold the child and the hosereel under the other. Like this we crawled back to the exit and out into the glaring bright sunlight.

As we dropped the dummies the BAI's picked them up and took them back into the building from the next team to search for.

My blue rubberised gloves were limp with the heat and I just had to whip my wrists to throw them off, mistake!

I then reached up to take my helmet off and nearly burnt my hands from the heat coming off of its shell. Was it really that hot?

We took our sets off and briefed the entry control officer of where we had been and what we had done so he could pass it onto the oncoming crews.

Whilst we were servicing our sets Graham and I dissected what we had just done and we couldn't really see that we had done anything seriously wrong. Cleaning done and the other teams were coming in and repeating what we had just done.

Drill complete and all the kit made up we went off for a cuppa and were told to wait in the mess until called for.

One by one we responded to Hemmings calls. Finally "Fireman Gough" echoed through the mess.

In the office was Stn O Moors, Sub O Ball and one of the BAI's. Unlike the abrupt Hemmings call the atmosphere in the office was completly different.

"Sit down Darran" said the Stn Officer.

"Well how do you think it went Darran?"

I thought I'd change my tack this time.

"OK I think Sir."

"Well you've passed you technical paper with 85% well done"

"Thank you Sir" I replied but still not sure if I should be happy or not as the practical result was still to come.

"I was in the crib room when you entered" Sub O ball continued.

"I was pleased you ducked under the smoke for a look but I was surprised you didn't see the casualty straight away, However your search pattern was systematic and once you found the casualty I was pleased you remembered the proximity search and came out with the chid as well. Well done."

I breathed an internal sigh of relief.

"Thank you Sir."

The BAI then chipped in.

"Your casualty handling on the way out left a little to be desired but you did get both casualties out reasonably quickly."

Stn O Moors then wrapped it up with.

"Darran, on the whole this has been a good week for you. You've passed and I'll be contacting your Officer in Charge to put you on the run as a BA wearer, only with an experienced partner, straight away, well done.

Now I could finally breath an external sigh.

"Thank you very much Sir."

"Off you go" was his reply.

As I left the office Hemmings was outside. Not a mutter of congratulations, so I just walked past him and back to the mess with an ever so slight grin on my face.

When we got back to station that evening we spent half an hour cleaning our kit before hanging it back on our pegs. As we were doing so John popped in on his way home from work and was very pleased that we both had passed. He already knew as he had received a call from Stn O Moors. He emphasied that we were only to wear with experienced BA wearers the LFF's and Nick, Tony and Simon until we had a few decent fires and wears under our belts.

That evening Graham and I found ourselves back at the station but this time in the social club where in the perverse way of the fire service (because we had passed our course) WE had to buy the drinks to celebrate! We were both extremly knackered and really didn't want to be out that evening, but secretly we were both hanging around for our first BA shout.

I am writing this during the 2020 COVID19 lockdown and have heard the news that Henry Moors has passed away. Not from COVID19. The very sad thing is that he has had to have an unattended funeral which is extremely unfair because there would have been a huge turnout from his old colleagues to give him a fitting send off. All I could do was to raise a glass to him in my home that evening.

CHAPTER 14
First Wear

Saturday night and I was fast asleep.

BEEP BEEEEEEEP BEEP BEEEEEEEEEP

Up like a shot, dressed and out the door. Moped starts first time and I'm off. It was about 0130 and the roads were quiet, so I was flying. I don't know why but at the bottom of Greys Road I turned left and went up Deanfield Road, not my usual route. Up to the top, turn right at George Harrisons then chicane it left and right to head down West Street and pull into the station.

Now because of where I lived, I usually only caught the second engine at night but as I dismounted the moped and ran into the station it was strangely quiet. B06L wasn't running and looking at the tally board only 3 tallies had gone and there was still a BA tally hanging there, my first BA tally and on the first engine as well. As I broke through the plastic curtains John said,

"House fire Deanfield Road persons reported".

"I've just come up there I didn't see anything" I said

I got my kit on and climbed into the middle of the cab where the BA sets were to find Rick was sat there.

"Alright Dazzzzzz" he slurred as he turned to me.

Bloody hell he was as pissed as a fart and was going to be my first BA partner!

Because the house was so close to the station we had to get completely dressed before we turned out. I leaned behind me and pulled the shoulder straps on and flicked the quick release buckle and the set fell forward onto my back. I slid forward and down off the bench seat, and this lifted the heavy cylinder up my back so I could tighten up the straps up.

By this time Tony was in the driver's seat and the engine was running. John was in the OICs and 2 more had climbed in the back, we were off.

In just over a minute, we were pulling up at the address. In the dark there was still no sign of fire or smoke, which would explain why I hadn't seen anything as I had raced past on my moped.

John had gone up to the front door and found the homeowner outside. He confirmed that everyone was out of the house which took the pressure off a bit. He said that the fire was downstairs, he didn't know what was on fire but there was an oxygen cylinder in the lounge, great! Pressure back on again.

While Rick and I were on the driveway starting up our BA sets, Tony was pulling off the hose reel for us and placing the branch at the front door.

As he came past, I whispered loudly

"Oi Tone."

When he turned to me, I just nodded my head in Rick's direction. He was clearly struggling with his BA set. Tony saw what the cause of this was and gave a look of desperation back to me. We both knew we couldn't say anything, now wasn't the time, but he whispered back,

"I've seen it. Watch your back we'll talk to John later."

John came over and told us he wanted us to enter through the front door with the hose reel and search the lounge.

We started up our sets and gave our tallies to Graham who was running the entry control.

All checks completed I waited for Rick to lead me to the door and take the hose reel as the experienced BA wearer. Oh no he literally pushed me forward in front of him to the door and gave me the hose reel to lead in!

I looked around to John for some signal that this was ok, but John was back at the pump, so I had no choice but to lead.

The front door was open, and I could see into the hallway leading to the lounge door which was made of glass but was shut. I could see that the lounge was heavily smoke logged.

I called to Rick to pull more hose reel through and when he was behind me, I opened the door. Once inside I put my face to the floor just as in training, but the smoke had reached the carpet and I couldn't see a thing, bugger.

I told Rick this but all he did was to push me further into the room. I started feeling around and found the coffee table, an armchair, then I caught sight of a faint flickering through the dense black smoke. I aimed the hose reel in the general direction of the glow and let rip. In a couple of seconds the glow disappeared, bingo!

I called Rick over, and we found that it was the far end of the sofa on fire which would explain the huge amount of thick acrid smoke.

Now the fire was out I grouped around and found the patio doors and opened them to start to clear the smoke. As it began to clear I caught sight of the oxygen cylinder in the corner of the room. I lifted it up and told Rick to get out.

As we got to the front door John was there to greet us. Rick just walked straight past him to his surprise, and I was left to tell John what had happened.

I left John with the cylinder and headed for Graham to get my tally back and shut down. Rick already had his set off and was climbing into the back of the engine. I just took my set off and sat on the drive to get my breath back. Tony came up to me.

"You ok Daz?"

"Yeah, no thanks to him, he's as pissed as a fart and bang out of order."

"I know I saw it. Leave it with me I'll have a word with John for you when we get back to station, don't you say anything".

"OK Cheers but I'm spitting bullets at the moment".

"I know you are and rightly so, but I know John and I know how to handle this."

"Thanks Tone."

I climbed into the cab and got a spare cylinder out and serviced my set on the driveway leaving Rick in the cab.

Once I'd finished the cylinder change and put the set back on the engine John came over to me.

"Well, that's your first wear under your belt Darran Everything ok?"

"Yeah fine" I replied somewhat unenthusiastically.

"What's the matter?"

"Tony will tell you later John."

"Arh OK. Now the room has cleared come in and show me what you found."

For the first time I could now see the complete room. It was much larger than I thought. I could see the sofa in the corner of the room, and I showed John where the flames were. We looked around the burns and found an ashtray in the ashes.

"Looks like we've found our cause," John said.

We brought the homeowner in, and he rather sheepishly admitted he had been having a smoke before he went to bed.

We left him trying to explain to his wife what had caused the fire and why.

Back at the station Rick hastily hung his kit up and didn't stay for the customary nocturnal cupper but signed the pay book and just disappeared.

Tony asked to see John in his office and as they went in the door was closed behind them.

People started to glance around and mouth what's going on?

I just mouthed back leave it.

After about 5 minutes the door opened, and John called me in and closed the door again.

"Darran, Tony has just told me what went on this evening with Rick, and I understand you are not happy is that right?"

"Not really John, no. I didn't have a chance to let you know at the time, but I wasn't happy it being my first wear and having to take the lead."

"I understand that. Is Rick still here?"

"No, he's already done one."

"Oh right. You are right to raise this and will you leave it with me to deal with."

"Yes fine, I just wanted you to know what happened"

"As I said I understand. On a different note, you did really well tonight taking everything into consideration. Can I ask you both to not discuss this outside of this office please."

We both agreed.

As we left the office the rest of the crew were finishing their tea and I could see that they were dying to ask what was going on, but I think they had a pretty good idea and kept quiet.

In the morning, I was eager to tell mum about my night. I could tell although she was interested, she also had a worry in her tone so anything else that I did I tried to play down a bit or just not tell her.

But I had my first operational BA wear under my belt, and I was buzzing.

At the end of the following drill night John called me into the office.

"I've had a word with Rick and told him in no uncertain terms he is in deep whatsit, but this time I'm keeping it on station. If it happens again, I will have no option but to inform division. Are you ok with that?"

I agreed. I didn't want to get Rick in any real trouble just to highlight the fact that he had scared the shit out of me, and I hadn't appreciated it.

I was always wary any other time I wore BA with him which is not good.

CHAPTER 15

HGV

One busy week down, one eventful weekend over and the next two weeks waiting. Monday morning, I reported to B10 Slade fire station in Oxford. Over the previous weeks I had gone through my medical and received my pocket sized cardboard folded provisional HGV licence, which together with my green paper ordinary licence I clutched in my slightly sweaty hands. I climbed the stairs alongside the pole drop to the top floor of the station and saw the tiny office with a brown and white Bakelite sign on the door "Brigade Driving School."

The door was slightly open, and I could hear voices inside. I knocked and was greeted with an instant "come in."

There were three people in the office as I entered. Two were in their 50s, one with grey hair and a bit of a paunch and the other looking rather scruffy with a mix of dark and grey hair tangled around his head. The third occupant was in his twenties, and I guessed a student like me.

"Are you Darran?" asked the scruffy one.

"Yes, morning"

"Hello there, I'm Colin and this is Maurice, and this is John, he's on his second week here."

I nodded, greeting them all.

"Is that your licence?" Colin asked looking at the papers clutched in my hand.

"Yes, here you are."

"Ok cheers. We've got some bits to do here before we get started so if you fancy a cuppa the kitchen is next door. Go and help yourself and I'll give you a shout in about half an hour."

Colin was an ex operational Sub Officer who had retired and then been taken on as the lead driving instructor rather than Maurice who was a civilian with no fire service experience. This explained why Colin was in a uniform with his old rank markings, but Maurice just had trousers and a jumper.

I took the opportunity to have a little wander around the station. It was a 2-pump station with one whole time pump, and one retained crewed pump.

The wholetime pump was out so the ground floor was empty.

With a nervous look around, I gingerly pushed the door release to the green double pole drop doors. As they opened, I was greeted by the shiny, highly polished by generations of firemen sliding down this sliver stainless steel fireman's pole. I quietly slipped through and carefully, quietly shut the two doors behind me.

I was now stood at the edge of the hole between the first and ground floors. With the doors shut behind me there was only one way down literally. With that I wrapped my arms around the pole followed by my legs and whoosh I was down, landing with a muffled thud on the cushioned pad on the ground, my first use of a fireman's pole. Being on a retained station we didn't have them, and I wanted to fulfil a childhood dream, and I had. Little did I know that just over 20 years later I would be using the poles every day once I had transferred to West Midlands Fire Service with lots of strange incidents shall we say.

I quickly scurried away from the pole, with a Cheshire cat grin, and back up the stairs to the top floor to make a cuppa and waited to be called.

A short while later Colin came into the kitchenette.

"Morning Darran, I'm Colin West and I'll be your driving instructor for the next two weeks, where are you from again?"

"B6 Henley Sir" I replied

"Ahh John Gosby's lot, good sound station there. And another thing it's Colin not Sir, or Sub or any of that crap unless there are white hats about, OK?"

"OK."

"Right, this first week we'll be doing some reversing exercises in the yard which on the test day you will have to pass. Then we'll take you out for some gentle driving out in the countryside so you can get used to the engine and the gears. There's no synchromesh so you will have to learn to double clutch. Have you done that before?"

"Erm no idea what that is".

"No worries. We'll go through it all before going out on the road."

With that Colin took me out to the garages at the rear of the drill yard. We slid open, or rather barged open the old faded red wooden folding doors to reveal an old Bedford TK fire engine. When the front-line machines had reached 10 years old, they were either taken off the run and used as spare machines for those that needed servicing or had broken down or sent to training or driving school.

This one was clean but was showing the faded red paint of years on the road. It had 2 large L plates with HGV under them. The front cab doors were the same as the ones at Henley but the rear cab doors didn't open outwards on hinges but were joined in two down the middle, so you pushed the middle, and they folded in together. It had no operational equipment on board, but the lockers were filled up with sandbags to simulate the all up weight of a fully kitted out engine.

The garage had a second engine in the bay next to mine with further bits and pieces stored further in, such as fire station open day signs and stocks for kids to throw sponges at people.

The back windows of the garage weren't clean and covered in spider webs and such like.

Colin then proceeded to show me the daily checks that the driver had to do. Oil, water, tyre pressures lights mirrors and finally horns.

Once these were complete, I climbed up into the cab and was shown what all the lights and switches were for and how to start the engine making sure the engine shut off knob was pushed right in and the vehicle was in neutral.

I gingerly turned the very small key in the ignition and the beast coughed a little first then roared into life, possibly because I had to accelerator pushed too far down, filling the garage with thick black smoke. Once things settled down, the smoke cleared, and I could again see out of the door, I engaged first gear and eased it out into the yard. Colin then explained all the controls and the manovering exercises we would be doing.

On the drill yard to the rear of the station were various painted yellow lines. These represented curves in the road and a garage that I had to reverse into then drive out of and go forwards into another painted garage without touching any of the lines. This would form the first part of my driving test in 2 weeks' time.

Using only the large side mirrors as there was no rear-view mirror, I began to turn the very large steering wheel, without crossing my hands, and slowly inched backwards around the curves and into the first painted garage. To pass I had to stop within 2 feet of the rear line without crossing over it. The first time I stopped 4 to 5 feet away then the second time I overshot the line.

"You having difficulty judging the distance Darran?" Colin shouted from the rear of the engine.

"You could say that" I replied.

"Here's a trick then, let me guide you back and I'll stop you a foot from the line."

With that he raised his right arm and beckoned be back until he turned his palm to me and shouted "STOP!"

"Now you are exactly a foot from the line. Can you see the bottom of the mud flap on the rear wheel in your mirror?"

"Yes" I replied.

"Well can you see the gap between the bottom of the mud flap and the yellow line."

"Yes."

"Well, if you reverse every time until the gap is the same between the mud flap and the line you will be spot on."

"Gotcha."

I then pulled forward and tried again, concentrating hard on that gap. When I thought it was right, I gently applied the brakes.

"Spot on Darran nice one."

We continued going backwards and forwards around the drill yard for another half hour or so until Colin called out, "Ok tea and toast time" With that I parked the fire engine up and climbed down rather pleased with myself as I had not hit any lines after the gap tip. With all the manovering in the yard I still hadn't been out of either first gear or reverse. When we got back to the mess the wholetime crew had returned and were already tucking into their tea and toast. Also, the cook was now on station, and there were no options I was told it was Toad in the Hole for lunch.

To have a cook on a fire station was the norm then. My current fire station 30 plus years later, Billesley in Birmingham, still has a cook but this is more of a luxury as due to the finance cuts when a cook retires or just leaves, they are not replaced now.

The cook handed me a plate with 2 pieces of thick cut white plastic bread toasted and smothered with butter and a white chipped mug of VERY strong tea. This wasn't really my cup of tea so to speak but I just nodded and said thank you. Colin took a seat at the far end away from the crew and beckoned me over to sit opposite him.

As I sat down, he pulled out a battered tin of Golden Virginia and flipped the lid open. Inside were several already, slightly tatty, rolled cigarettes. He pulled one out from amidst the loose tobacco and flipped the metal cap off a petrol lighter and proceeded to light this small "fag" with a flame 3 inches high from the lighter.

Now with the tin put firmly back in his pocket he began to explain what double de-clutching was.

In most cars there is a synchromesh which keeps the engine and the gears spinning at a constant speed when pressing the clutch pedal and changing the gears up or down. These fire engines didn't have this. They had what was called a crash gear box, so when driving and changing gear you had to depress the clutch pedal, bring the gear lever back into the neutral position, release the clutch then rev the engine and whilst the engine revs were still high depress the clutch again and the push the gear lever into the gear required. This sounded really complicated to me, but I was told that we would have a go in the yard using 1st and 2nd gears before we went onto the roads.

We duly finished our tea and toast and gave the dirty crockery to the cook and headed back to the drill yard and our fire engine.

Now back in the cab with the engine rumbling underneath me, and Colin sat in the front seat I pulled off in first gear. Halfway across the yard Colin said,

"Right now, change up."

With this I depressed the clutch and pulled the lever back

into neutral and then my rather heavy foot took over and to quote Colin afterwards,

"Revved the bollocks off it"

Whilst the revs were still far too high, I depressed the clutch again and accompanied by a loud graunching noise, I repeatedly tried to pull the gear lever back into second gear. It did not want to engage and Colin had to shout out above the noise of the gears "BRAKE!" as the concrete wall at the far side of the yard was getting mighty big.

We came to an abrupt halt, inches from the wall and I gingerly applied the compressed air hand brake.

Colin didn't say a word he just got his fag tin out of his pocket and promptly lit another rollup.

"Right, let's back up and try it again with a little less revs, ok?" he said in a slight sarcastic and relieved tone.

After a dozen or so runs across the yard the kangarooing and graunching calmed down and I began to get the hang of it.

Many, many years later when I was attending one of my 3 yearly driving refresher courses in the now modern Volvo fire engines with a manual gear box BUT now with synchromesh, my instructor, who was new to the brigade commented,

"You're double clutching, it's very smooth but you do know you don't need to, don't you?"

"Yes, but it was the way I was taught many years ago and I now can't get out of the habit."

Colin now popped the tin out again and lighting another rollup and after igniting it turned to me.

"Right then Darran lets venture out of the safety of the yard. Go through the gate and turn right towards the ring road. Just remember that the acceleration in this is not like your car and you need to allow far more distance from oncoming cars, and they come down the road outside just a bit quick"

I edged out of the yard and up to the main road. The cars were, as he said, rather quick, and it seemed like an age before I had a gap both ways big enough to pull out, and this was only because a car from my right actually stopped and flashed his lights at me to pull out.

Colin said that I was to be careful as many drivers will think they are doing the right thing and stop or give way to you, but you may find times that you will be waved into situations where you don't really want to go, and that is still true today.

I trundled on down and joined the Oxford Ring Road which was dual carriageway and spent an hour or so just going around it changing gears and negotiating the many roundabouts.

Lunch time was approaching, and we returned to the station and found our way up to the mess and were promptly handed our plates of Toad in the Hole with veg.

I was met with a mass of batter a couple of inches thick with some fat pork sausages poking out surrounded by a mass of carrots, cabbage and beans boiled to within an inch of their lives!

I would describe it as mass industrial cooking rather than homely but by then I was starving and eventually finished the lot.

After my meal the crew had disappeared. Slade station had a large social club with a bar adjoining the mess and off that were two doors. One led to the fire escape but the other led to a TV room with comfy chairs.

I gingerly poked my head through the door and was waved into a seat. It was half past one and a new soap was just launching on the BBC called Neighbours! This was how I got my introduction to Kylie and co.

After lunch we were back out on the roads and this time we went off around the countryside and down some rather tight lanes. This taught me to use the extra height I had in the cab and to look over the hedge lines to see further ahead for oncoming traffic and to anticipate if I would have to stop or not.

At one point Colin took me through a small village called Islip. I still regularly drive through the village now and always remember this first encounter. It is a very old village with many stone build houses and narrow roads. It also has a one-way system so once you enter it you can only go forward.

As I rounded a tight corner with me having the right of way Colin said

"You go down here to the right and as you can see it is tight."

He wasn't joking either. It was down a hill between two stone houses. The walls were already scarred by drivers who had misjudged their approach.

As I eased past the corners of the buildings, I must have breathed a sigh of relief as both the wing mirrors had missed the walls by 4 or 5 inches.

"I wouldn't relax just yet" Colin whispered "it gets tighter"

Just as he said this, I saw 2 more walls leaning into the road with white paint on the corners.

You're joking how the hell am I going to get through that? I thought.

"You will only have an inch or so from each mirror so align your drivers mirror first with the edge of the wall and the nearside mirror will pass through." Colin advised.

I slowed right down and with the offside wing mirror brushing the ivy which was hiding part of the wall I squeezed through.

It reminds me of a tongue in cheek phrase we now use.

"The faster you go at a gap the wider it gets!"

"Now you can breathe," Colin giggled.

We spent the rest of the afternoon exploring the Oxfordshire countryside and drew back to the yard around 4 ish.

After Colin had shown me how to refuel the appliance and record the diesel issue in the logbooks, he saw me reverse back into the bottom garages and "put her to bed for the night"

"Right how do you feel that went?" Colin asked as we walked back across the yard to the station.

"Bit rough at the start but I really enjoyed it, I'm knackered now" I replied.

"That's about right. Well, you get off now and I'll see you tomorrow."

With that I went and got my bag and headed off to my car. Much to my embarrassment once in my car I firstly stalled it then over revved it in the yard. I had been so used to the heavy pedals of the fire engine that when I got back into my own light car it took a minute or two to reacquaint myself with its much lighter controls.

Once home Mum gave me a proper home cooked meal and I hit the sack early as I was absolutely shattered. Fortunately, I had no fire calls overnight and the following days were again spent driving around the countryside.

When I went to drill on Tuesday night everyone was asking how it was going and who was my instructor. When I said it was Colin there was a unanimous seal of approval for him. But more than one gave me a tip about him. I was asked did he smoke much to which I replied like a chimney. The tip was to watch how many he smoked. It was said that as a person's driving got better, he smoked less and less. I observed his smoking habits closely after that and it would appear that they were right.

When we got to Friday, I was allowed to watch John's driving and reversing test in the yard from the mess room. It didn't look too bad. The examiner was also a brigade officer Station Officer Ade Wright. He was the Station Commander at Slade and had qualified as a Dept. of Transport driving examiner, so we didn't have to book into the civilian test centre.

We spent the rest of the morning driving around the city and would be going into the built-up areas next week.

We were back in the yard by lunchtime, and I was told to wash and fuel the engine then put it back in the garage. This I did and was told I could take an early day and go home.

I had enjoyed the week so much but my legs, especially my left leg, clutch foot, was aching like hell.

Week two started much the same as week one only now I wasn't the newbie and was chatting in the office with Colin when the latest recruit arrived.

Colin didn't waste any time. We were straight out on the road but this time going through the Oxford city centre and talk about having to have eyes in the back of your head! Driving through the ancient narrow congested streets fighting with the hundreds of busses was bad enough but being a university city, I also had to contend with the thousands of ninja style cyclists who had no care whatsoever for their own safety or my lack of instant moveability in a big red lorry! It was a nightmare, Colin had warned me beforehand, but it was ten times worse.

I began to wonder how the city fire crews dealt with this bedlam especially when driving to a shout on blues. Respect.

After an hour we escaped the city and headed for my own

station ground of Henley and the surrounding area. We called in at the station and I raided the tea cupboard for us.

During this week I had surreptitiously been keeping an eye on Colin's smoking and I was extremely happy because he had stopped smoking in the cab now and only had one when we stopped for a break. Whether this meant that my driving was getting better, or he was so scared whilst I was driving that he was frozen in the seat, I prefer to think of the former.

All too quickly it was the Friday morning of my test. I had spent the previous night reading and re-reading the highway code for HGVs and was fairly happy with it.

There was no written test in those days, but I knew I would be asked some questions during the test about the differences of HGV driving.

Ade Wright came into the office and following some pleasantries and a little bit of micky taking from Colin about my driving I left with him to go to my engine. I took him around the vehicle showing that I knew how to do the daily checks that the fire service required as well as the standard safety checks required by the D.O.T. (Department of Transport).

With that completed I started up the engine and positioned myself at the start of the painted line obstacle course. When he gave me the signal to start, I began to reverse around the slalom course and into the garage. Even though I had done this every day of the two-week course I was still nervously watching the bottom of the rear wheel mud flap to stop with the appropriate gap between that and the painted line on the yard. When I thought it was about right, I gently braked and Ade immediately began to walk to the passenger seat. Bingo I'd got it first time.

We then headed out onto the road. Ade had a nice calming

voice and the directions he gave were always with plenty of time for me to position the engine and take the correct line into hazards and bends. A couple of times in the countryside I did slice a corner or two and clip the verge but on the whole, I was happy with my drive as we pulled up back in the yard at Slade.

I was asked a few questions about HGV driving that I got mostly right and then without indicating any result we went back upstairs to the instructor's office.

Ade began "Well Colin the nearside does need a bit of cleaning the number of muddy verges we clipped and there are a few cyclists with grazed elbows in the high street"

I must have had a shocked look of disbelief on my face as Colin glanced at me and gave the disapproving parent look at their child who had misbehaved at school.

After a pause they couldn't hold it any longer and burst out laughing.

"Your face was a picture" Colin said

"Well done you've passed but just take a slightly wider line into the bends in the country" Ade added grinning.

I've told him about that" Colin added also grinning.

With that I gave them my provisional licence to send off and in return I received a logbook. In this I had to clock up 150 miles or six months of driving on the appliances at the station before I could drive on Blues and twos.

This is greatly different from the procedures nowadays. You still must complete a two-week initial driving course, but it's shared between two students so in effect you only get 1 solid week of driving. Once you pass then you still must accumulate the 150 miles or 6 months driving but after that you must attend another course of a week to be trained to drive under emergency conditions.

This course is entitled the EFAD or Emergency Fire Appliance Driver course, or rather as we call it Extremely Fast And Dangerous.

This course teaches you road craft and how to anticipate dangers and to be assertive, not aggressive, with your driving as hesitation can cause more accidents than being confident with your approach.

You also now must re-qualify every three years on all vehicles you may drive.

On one of my requal. days a while ago I was asked to complete an electronic survey for a university student about driving skills.

One question was "Do you think you are a better driver than most drivers on the road?" to which I answered yes.

The automatic results this survey generated said that I was a danger on the roads due to over confidence! When the instructor came to me with these results and asked me my opinion I stood by my answer. He asked me to explain this, so I did.

"Firstly, when someone passes their ordinary driving test that usually will be the only test they will take in their whole driving career. I on the other hand have been professionally trained to drive an HGV at high speed through every type of road condition. I am also re-tested on my skills every 3 years. After 50 years old I have regular medicals to ensure I am still safe to drive. I also have taken and passed the Institute of Advanced Motorists test."

"OK stop there you are right and actually we completely disagree with this survey anyway."

As I write this chapter, sat in my fire station over the Christmas break, I can rejoice that at nearly 55 years old I have just passed my HGV medical again with flying colours to renew for another 5 years. I also passed my 3 yearly EFAD with a lovely comment from my examiner, a lady who had just transferred to the fire service from the army where she was also a driving instructor. We had spent the day on blue light runs in both a large Volvo fire appliance and a Toyota 4x4 Hilux which is my usual vehicle and her end of course comment was.

"I felt very safe on all the drives today."

I can't really ask for more.

On a sad note, though I have just heard that Colin West has just passed away. The Facebook page for past members of Oxfordshire Fire service has been flooded with lovely comments and memories of Colin and well deserved to. Rest in peace Colin.

CHAPTER 16
Back to Earth with a Bump

Back at the station that Friday evening up the club everyone was congratulating me on passing my HGV. John told me that although I had passed, I would have to wait until it appeared in routine orders before I could start clocking up my miles driving back from shouts.

The weekend passed uneventfully, and Monday morning arrived and my first day at the china shop.

I rode my moped down to the station and left it inside for safety and walked down through the town to the shop opposite the church, and more significantly the pedestrian crossing! As I have previously explained the bleeping from the crossing when someone had pushed the button was almost identical to my alerter. Throughout my time at the shop, I was always jumping when someone was crossing the road.

The shop was very old fashioned. It had 3 stone steps leading up through the door which had a bell on a curved piece of metal to jingle when a customer entered. It was actually two shops joined together through a small archway inside. One shop sold china and ornaments of all kinds whilst through the archway were very upmarket ladies accessories, handbags, scarves and jewellery etc.

The main china shop had two internal arches and below one was a carpeted trap door leading to the cellar. In here were kept the empty boxes for the items on show. Next door also had a cellar, but this was accessed by a door and very tight and steep wooden staircase. Here were duplicate stock holdings and yet more boxes.

This cellar had a musty damp smell which wasn't surprising as with the river Thames only a stone's throw down the high street, we were very near to the water table level. What I loved down there was the fact that the whole building had been constructed from recycled Thames barges. The solid oak beams that had been reused still had the original cuts and fixings from when the plied their trade many years ago along the river. I've always been a huge fan of history, which was seriously fuelled by my history teacher Mrs Pam Syrett and form teacher, also a history teacher Mr Bob Phillips, both of whom I remained in touch with, in my adult years. Both have now sadly passed. I was able to go to Mrs Syrett's funeral, and it is testament to her popularity that a large number of ex pupils turned up to say goodbye to her.

The first thing I was shown was where to make the tea and who had what. Get the important things over first.

As the morning was very quiet, we sat down and Mr Hickman, Bob, asked me how my previous three weeks had gone. This was a family business, Mr Hickman was a tall elderly man in his early 60s with straggly and thinning grey hair. He was somewhat overweight and almost always wore a brown leather jacket. He had a way of showing interest in what you had been doing to pass conversation, but I could see he very quickly would get bored, and you would lose his attention very easily. His wife was a lot smaller with a bustling attitude always walking around with a duster looking for something to dust.

They owned both properties and their son Bryan lived in the flats above the shops with his wife and young daughter. Bryan was very much the businessman of the family and did the majority of the buying and promoting with his wife attending to the bookkeeping. I got the impression that he looked on his mother and father as being more of a retirement couple passing the time in the shop, I think Mr Hickman senior had different ideas.

Behind the counter was an annexe which led out to the long garden for the flat above. The annexe was where most of the packing was done for their very busy mail order and export side of the business.

There was also a second china shop around the corner in Duke Street. This was run by two ladies Jennie and Ruth. It sold the same china goods and had the same style of cellar as well.

The morning was mainly spent with Bob showing me around the shop and introducing me to the hundreds of different makes of china and ornaments for sale. This is where I got my liking for Coalport, Border Fine Arts, Lilliput Lane, Wedgewood, Port Merion and far too many more to list here.

Lunchtime arrived and I made my way up to the Row Barge pub behind the fire station for a bite to eat. This became a regular event most days. I only had half an hour and so phoned ahead to order my food and it would be hot and waiting on the bar by the time I had made the 5 min walk up the hill.

That afternoon I was in the annexe learning how to complete the export and VAT exempt paperwork with Bryan when my alerter went off. I jumped, Bryan jumped, and I began to move, not at a run, but faster than I sensibly should in a china shop, Bull in…. you get what I mean. As I reached the door a tall man was just about to enter, I ducked and tried to squeeze past him, but I didn't duck low enough, and head butted him in the mouth. I shouted sorry and ran off up the road to the station. The shout took us back down Hart Street past the shop and I glanced across to see the man in an animated discussion with Bob in the doorway. Bob waved to me, and the man turned around to see us passing.

When I got back to the shop an hour or so later Bob called me over. Oh, great I thought what a start to my first day. Assault a customer and get a rollicking.

"Darran" Bob calmly addressed me with an expressionless face.

"When you get a fire call please walk calmly out of the door before running off and try not to assault big Australians into the bargain!"

"I'm very sorry. Was he ok"

"He came in and asked me if I had seen what had just happened while he was wiping blood from his cut lip? I said I had."

At this point Bobs face cracked into a smile and then developed into full blown laughter.

"I said it was his fault for getting in your way as you were running off to a fire call, and when he saw you in the fire engine, he was fine. Just walk in future please".

"Sorry Mr Hickman I will in future".

"You did brighten our afternoon I must admit, and it's Bob".

The next few days were taken up with learning the ropes in the shop interspersed with fire calls.

The first experience of responding in the shop had calmed me down a bit but I still had a sense of panic when, holding a large box of Wedgewood china, that I was about to start wrapping up for export, (about £2000 worth), and gingerly making my way up the stairs from the cellar in the shop with my head just appearing through the floor, that my alerted went off!

I shouted, "Help will someone take this off me please?"

I was rescued by Bryan, and I was off.

Being put in charge of the export packing and posting did have its advantages. I was the one to take the parcels down to the post office.

Henley post office was a cathedral type building with its cavernous walls covered in oak panelling and they still had pay phones inside wooden cubicles with bi-folding doors. The queue used to wind itself around the floor waiting for your turn at the full length of the room counter.

There were usually 3 or 4 people behind the counters and when I reached the front of the queue many a lady behind me became very puzzled and grateful when the next counter became free and I would let them go ahead of me, even though I was holding a huge box. I had an ulterior motive behind my good manners, and her name was Anna.

Anna was a lovely girl about 5 foot and a couple of years younger than me with jet black hair and a smile to light up any room.

I would always try to time my progress in the queue with the speed of exit of the previous customers. Later, as our friendship developed, Anna would also try to speed up or slow down her service to coincide with my arrival at the front of the queue.

As we chatted over the parcels it turned out she was the sister of one of my old primary school friends. I found out when her midweek day off was and arranged with Bob to have mine on the same day.

This was a lovely warm September day and after I had dropped off the usual parcels with Anna, I had a spring in my step as I walked back to the Hart Street shop. Once back in the shop Bob asked me to pop back to the Duke Street shop for an item of stock, this was quite usual going backwards and forwards between the shops.

As I got to the corner, I heard a scream and the hissing of lorry brakes. When I turned into Duke Street I was met with an horrific scene. The lorry had stopped across the traffic light line and there were people screaming and staring at the wheels where I could see the lower body of a girl under the rear axle of this earth moving lorry. It was immediately obvious to me that the girl was dead.

As I got closer my alerter went off and I had to turn around and run up to the station. As I got to the doors, I heard John call out "RTC Duke street persons trapped"

When I heard this, I immediately said, "John she's not trapped she's dead I've just come from there" and I told him what I had seen.

We quickly got dressed in our fire kit and both engines went the very short distance to the scene.

John had a look, and it was obvious that the poor girl was dead. I won't go into any details out of respect.

Our first job was to get the large green salvage sheets from the engines and string them up across the street to shield the horrific scene from the public.

We got the lorry driver out of the cab and quietly sat him

down quietly out of the way of the activity. He just kept saying,

"I didn't know anything about it until people ran out in front of me shouting and waving for me to stop. I didn't know, I didn't know, I didn't see her at all."

There was nothing we could do at that moment. The girl wasn't trapped because the rear wheels had rolled over her and stopped just past her. All we could really do at that time was to shield the area and wait for the police scenes of crime officer to arrive and record the scene for further investigations.

John told me to go and stand at the cab end of the lorry and keep everyone from passing or gawping.

After doing this for about 10 minutes and getting rather tired of repeating myself to everyone who came up to me a man came running up to me and as I stepped into his path and before I had a chance to repeat myself yet again, he body charged me and sent me flying into the road.

"HEY, STOP" I shouted as he attempted to break through the canvas sheet screen.

At this commotion two police officers appeared from the other side of the screen and following a little struggle calmed him down and forcefully but carefully walked him back past me and sat him down in their car away from the scene.

John came around the screen and saw me prostrate on the road.

"Darran are you ok?" he asked

"Yeah, thanks John What the bloody hell is his problem?"

John quietly whispered "He's her father"

My heart sank, and seeing the obvious look on my face John took me back to the other side of the canvas wall out of sight.

Now we were just stood around for an hour or more just looking at the grisly and very distressing scene. I popped back to the shop which was only a couple of doors down the road and asked Jenny to call Bob in the main shop and tell him that I was going to be some time. The butchers next door then offered us all a cuppa which we all readily accepted.

Now this was the first time I had experienced the emergency services "black humour".

Maybe it's a pressure release thing, I'm not sure, but it's the sort of humour that is only really understood by those in the services and most certainly must NEVER be heard by the public. At the very least it would be seen as "in poor taste" or "disrespectful" or just downright "out of order" and in these enlightened times would most certainly end up in a complaint at the very least.

But with what we were looking at and the tension palpably building one of the coppers whispered something, which I will not repeat, but it made us all look at each other and firstly with a look of "did he really just say that" followed by an extreme effort to not either laugh out loud or make any reaction that could be heard or seen by anyone else. But it was probably the right thing to say at the right time to relieve the building pressure, but it was, to almost quote Kenny Everett, "It was done in the best possible BAD taste."

Eventually the scenes of crime officer arrived and took many photographs of the scene. Once he was finished, we were asked to help with the removal of the girl which was carried out with extreme respect.

I suggested to John that I could go to the butchers and ask for some sawdust to help clear the road and he agreed. This was probably one of the most distressing things I have, and I think, ever had to do at a job, I'll leave it there as it's too painful even 35 years later to record here.

Once the lorry was moved away by the police, we used our hose reels to wash down the road and the sides of the shops.

We then quietly boarded the engines and returned to the station. The cab was strangely silent and then this was the only time I experienced this whilst stationed at Henley.

The usual practice on return to station was to immediately re stock the engines with any consumables we had used and to then wash them down, but this time John told us to stop and go upstairs and open the bar up.

This was absolutely what we needed and is nowadays formally recognised as "Critical Incident Defusing" but at that time it was just bloody good management from John.

We all poured a pint and sat down. John asked us if we were ok? which drew a series of quiet nods.

He had been talking to the police and it appears that the girl was walking along the path towards the traffic lights when they changed to green. It's still not certain even now if she slipped on the curb as the street there is very narrow, or she fainted in the heat, but as the lorry pulled away, she fell under the double axels of the heavy earth mover and the result was instant.

As I said the driver was no way at fault and didn't feel anything. The first he knew was people running out in front of him screaming and waving at him to stop.

Eventually I got back to the shop and obviously the word had got back as to what had happened.

I must have looked pretty rough and when Bob saw me, he

immediately told me to get my stuff and go home. I appreciated this as I really didn't need an interrogation as to what had happened.

They say that some instances stick with you for life, and this is one that is still vivid in my memory with images I will never forget.

CHAPTER 17
First Drives

Now that my name had appeared in Routine Orders, I could start to drive the fire engines back from shouts.

Even though the driving school engines were weighted with ballast to simulate the all up weight of a fully operational truck, your own station trucks were different, and to get used to the vehicles a newly qualified driver had to clock up 6 months of normal driving first before graduating to blues and twos.

The first few fire calls I got on as we were getting ready to return to station, I was asked by John if I wanted to drive back. I was furious with myself because I couldn't for the simple reason, I had left my brigade issue shoes back at the station by my kit peg, and you were only allowed to drive in brigade issue shoes.

Finally on a Friday morning we got a call to an address just outside of Reading. It was still in our ground but probably the furthest you could get away from our station. As I was grabbing my fire kit from my peg John called out

"Darran don't forget your shoes."

I quickly stuffed them under my arm with all my other gear and clambered onto the fire engine.

The drive out was uneventful, and the actual call was a false alarm, but for the whole time I could feel my nerves building as I knew what was to come.

As John was walking back to the engine, he signalled to Tony to vacate the driver's seat and for me to undress from my fire kit and swap with him.

I left all my kit in the rear cab and calmly, on the outside, but nervous as hell on the inside, climbed up into the driver's seat. This was the first time I had driven an engine with more than one passenger in it and I could already feel the 4 pairs of eyes burning down on me from the rear.

"You know the way back Darran, off you go then" John said nonchalantly.

I eased out of the farmyard and back onto the road leading to Sonning Common.

I drove through the village safely and negotiated the two hills with sharp bends successfully. My confidence was growing, which was to be my downfall.

As I approached the next turning, known as Bolts Cross, it was downhill with a blind bend to the left, and I had to turn right at the bottom of the dip.

As I went down the hill, a bit too fast, I indicated to turn right and instead of slowing right down to see around the bend I drifted out and as I was level with the turning, I turned.

As I did a car came around the bend and didn't have to brake sharply but its nose did dip. I heard the exclamations and sharp intakes of breath from behind me. Once around the bend I summoned up the courage to glance across at John. The look said it all, I was in trouble.

The remainder of the drive was a lot slower than it needed to be. I reversed on the drill yard with someone seeing me back into the station. I turned off the engine and completed the vehicle logbook.

The atmosphere was weird, there was chat, but none included me.

I quietly hung my kit back onto my peg and as my helmet clunked onto its hook, "Darran will you come into the office please" came the quiet call from John.

I was expecting this and quickly jumped the 3 steps out of the muster bay and stood in his office doorway.

"Come in and close the door" I was instructed to do.

I couldn't look him in the face, I knew he was not happy to say the least, but I just thought, take it and don't argue.

"Darran" he began, "I don't for one minute believe you drove like that on your driving course, did you?"

"No" I meekly replied.

"I don't believe you were showing off either, but if you ever drive like that again I will take you off driving, is that clear?"

"Yes Sir," I quietly replied.

"Off you go."

I left the office. It had taken less than a minute, but boy it had had the desired effect, I had been well and truly bollocked.

For the next few months, I diligently drove back from shouts in both B06L and B06W they both handled slightly differently but I was enjoying myself.

In those days once you had completed your 6 months and your station officer was happy then you would automatically go onto blue light driving.

Nowadays you must complete a weeklong EFAD course. Emergency Fire Appliance Driver or, as we call it, Extremely Fast And Dangerous. You also must complete a 3 yearly refresher on blues and twos for each vehicle you are passed to drive to emergencies. For me now that's 4 vehicles. A PRL Pump Rescue Ladder (a fire engine to you and me) A BRV (Brigade Response Vehicle) which is a Toyota 4X4 Hilux, my normal vehicle now. The ICU, Incident Control Unit, affectionally known as the "Horse Box". This is the mobile control unit for large incidents. Finally, the CSV Command Support Vehicle, a smaller control vehicle which is a long wheelbase Transit van.

But back in 1987 it was a case of John saying OK you can drive on blues now, and that was with no practice just straight in. When John told me and the rest of the officers that on a drill night, I was champing at the bit for my first drive.

Over the next couple of days shouts came in but were either one pump calls which I missed, or I was too slow for the drivers tally and no one else would give up their drivers tally for me, oh no.

Finally, one Saturday afternoon I got my hands on that treasured red driver's tally. It was for the second engine B06W and I was happy as Larry until I heard what the incident was.

I was hoping for a little alarm actuating or a small grass fire or similar for my first Blues & Twos, but oh no, Goughy has to do everything in style. No piddling little job for you my boy, no you can have an aircraft making a forced landing over the river in Berkshire!

So I threw my kit into a rear locker and pulled it shut. I climbed into the cab and turned the key. The diesel warming light went out, I turned it the next notch and the cold engine erupted in a cloud of blue smoke rapidly filling the appliance bay.

"All in?" I shouted over the noise of the engine, and as Mick Goddard climbed into the officer's seat he said, "Right then Darran do you know where you are going?"

"Yes "I replied, I had heard John saying that he was going to go up Remenham Hill and in from the top. "Do you want me to go down by the Cricketers pub instead?"

"Yes, good idea approach from both ends" Mick confirmed.

I pressed the clutch pedal down, pushed the gear lever into first and slipped the handbrake off and eased out of the station.

With a grin on his face Mick asked, "are you going to put the blues on or what?"

Bugger I'd completely forgotten to switch the lights and sirens on.

"Oh, I suppose," I replied grinning back. Once the two tones switch was on, they were operated by the drivers left foot on a push button alongside the clutch pedal.

As I pulled out onto West Street, I hit the button to give plenty of warning for traffic on the blind junction at the bottom of the road.

Working my way through the town centre, now I can look back and laugh, but it must have looked like we had Kangaroo diesel in the engine. I was so nervous approaching the traffic lights that I was finding it hard to keep my left leg still on the clutch pedal, it was shaking so much that the engine was lurching up and down all the way through the town centre.

I finally got over Henley bridge and turned onto the lane we wanted and although it was a bit twisty, I could get some speed up and smooth the ride a little.

As it turned out we got to the aircraft first. It had made a perfect landing in a field alongside the river and the pilot was climbing out of the cockpit. It was a little single seater propeller driven German aircraft. I pulled up alongside it and applied the hand brake, put it into neutral and let out a sigh of relief. First Blues and Twos done safely, just a bit jerky.

B06L arrive a minute or so later, head on to me as we had taken different routes to locate the plane.

John walked over to the pilot and confirmed all was well and it had just been an engine failure and that no one was hurt.

He then started to dictate a stop message for me to send back to fire control over the radio. It was the driver's job as well as driving, operating the pump, and finding the nearest fire hydrant, to send the radio messages.

John started,

"From Station Officer Gosby at field adjacent to the River Thames, Berkshire. Incident is a single engine.............."

He paused then on the radio asked "what sort of plane is it?" I quickly replied "Focker John."

There was another pause then John answered.

"Darran, I know you've just had your first blues drive but there's no need to talk to me like that!"

I quickly replied "No John It a German aircraft manufactured by Focker that's F O C K E R"

"Oh right" he answered. As I looked over the field, I could see the rest of the crew doubling up in stitches of laughter, at John or me. I'm not quite sure.

The drive home was a little smoother and I reversed back into the appliance bay nice and straight.

To this day, 37 years later as I write this, I'm still driving and love every minute of it. There are many different challenges driving in Birmingham city centre to Henley but that makes it all the more challenging.

CHAPTER 18
Evel Knievel I'm Not

Now, as I've mentioned before, I used to respond to shouts on a 50cc moped after having to return my company Ford Escort when I left the photographers' studio and started working in Henley.

It wasn't a particularly fast moped, but for getting through traffic quickly and mostly legally, it was superb. When responding at night from home, I had to negotiate 2 roundabouts enroute to the station. The first, on my estate, was a left turn, no problem there. The second, one a hundred yards later, was a right turn onto the main Greys Road. In daytime, I would take this roundabout properly and go clockwise around it. At night, I would look out for headlights as I approached it, and if the road was still dark when I got to it, I would cut the corner and do an immediate right turn and then carry on.

At night, I would often be on a kind of automatic pilot when responding (half asleep). Frequently I would wake up as I approached this second roundabout. Maybe it was because subconsciously I knew that I had to pay attention on this turn. I would think "Oh I must be on a shout" and then I was ok for the rest of the journey to the station.

As I got to the town centre, I would turn into Greys Road car park and cut through it, coming out at the town hall. Then I would turn left up Gravel Hill and then cut through another car park in front of the Henley Exhibition Centre, which was the old fire station. Then, at the end there was a footpath running upwards alongside the end of the drill yard. This was wide enough to get the moped up. At the top, before I turned towards the station, I had to check that no one else was coming into the yard in their car. I would cut across and step through the middle of the moped onto one footrest and jump off, push the stand down, take the keys out and run into the station. That's how it would go when it went to plan!

A couple of occasions, it didn't go according to plan. It was a winter's evening about 1730 just before a drill night when the alerter went off. I was at home in uniform ready to leave for the station, so I grabbed my helmet and off I went on my trusty moped.

As I reached the entrance to Greys Road car park, I met Graham Case on his bicycle coming from the opposite direction. This often happened and we would cross each other's path at the start of the car park. I would, as on this night, go the wrong way through the car park, usually attracting beeps from horns or flashing headlights, whilst Graham would go the right way through. We had two rows of parked cars between us then we would merge at the far end of the car park heading towards the town hall.

This night, as we merged, I hit a pothole in the road and my front wheel rose into the air. As it did, I gripped the handlebars and held on for grim life. Unfortunately, this was not a good thing to do, for as I held on to the handlebars, I was also holding the throttle fully open!

The front wheel reared up in front of me and I slid off the back and was standing holding the moped, like a raging stallion in a rodeo, in mid-air and spinning around in circles. When the realisation hit me that I needed to let go of the throttle, I released my right hand and the screaming tone of the engine died and the front wheel started to drop as I continued to spin around in a circle.

Unfortunately, as the wheel came down, it swung right into Graham as he was at full peddle to my left. There was a resounding thud and we both ended up in a heap of bicycle and moped in the middle of the road surrounded my stunned car drivers.

I looked at Graham and he looked at me, "You Ok?" I asked "yeah, you ok?" Graham replied "Yeah, let's get out of here" I replied, and we picked ourselves up, Graham remounted his bike, and I picked up and restarted my moped and we zoomed of laughing our heads off, heading towards the station. As we got to the footpath B06L was already halfway down the road and we could tell it was only a one pump shout as B06W was stood silent in the station.

It turned out that the call was only to a fallen pregnant horse which couldn't get up!

One Sunday afternoon, I was responding to a call and did my usual route on the moped. As I came up the footpath at the bottom of the drill yard, B06L was just pulling out. I did a sharp left turn and, as I brought my left foot through the gap in the moped to slow down, I must have hit some gravel and the wheels went from under me and I went flying across the yard right in front of B06L. Very embarrassed, I found my feet and pulled the moped out of the way, to hear the raucous laughter from the rear cab fading as the engine went down the road.

I walked into the station and parked my moped up. Only then did I notice I had gravel rash down my arm and hand, and it was beginning to sting somewhat.

It was only a one pump shout to Shiplake, so I went into the washroom and started to bathe my arm and gently wash the gravel out of the grazes.

Whilst I was the recipient of some "fallen off again" and "What the hell have you done" comments, it was light hearted.

As the last bits of dirt were washed from my arms, we heard John on the radio, "Make pumps two, boat house well alight" we sprang into action well before the station bells went down. My arm was as sore as hell and was weeping a bit, but I had the driver's tally and wasn't going to give it up.

As we headed out of town, I was working out the best way to the address. The house was at the end of a long road but there was a short cut down a very rough road. I was familiar with this house, because it had a massive garden railway and, when as a kid living at Shiplake lock, I did every paper round as well as both the milk round and the butchers round in Shiplake. So, I knew the village well.

Mick was the officer in charge, and he told me to take the unmade road. However, for two reasons I said no. Firstly, it was extremely rough, and the engine would suffer, but secondly the railway bridge halfway down had a Three-ton weight restriction on it. We could see the smoke in the distance but took the slightly longer but faster route through the village and down past the railway station. We were praying that a train wasn't due, and the level crossing would be clear, this would hold us up even more.

As we approached the river along the driveway, sure enough, there was the garden railway. John called us on the radio and instructed us to get the lightweight portable pump, (LPP), set into the river and working asap, as they were out of water.

We heaved the pump off from its stowage cradle and piled on top of it all the other equipment we would need. Then four of us struggled to carry the load the 50 yards or so to the river. Nowadays, as a health and safety rep, I would have a fit at the weight we carried but needs must.

We set it down on the grassy riverbank and connected the hard suction up and dropped it into the River Thames. We ran the hose from the pump to the fire. Bullet Trendall, started the pump, and we had water very quickly. We left Bullet to look after the pump.

Bullet was a lovely old guy. He was an HGV skip lorry driver for Grundon's Waste Disposal and had a heart of gold. He was very quiet but was a laugh in the social club at night. He was close to retirement, and we tended to give him the less energetic tasks while us young bucks did the running about.

If you've seen the film with Steve Martin "Roxanne" about a fire department chief with a big nose, then you will have seen a character called Andy who is played by Michale J Pollard, and he is Bullet to a tee. In one scene, his fire coat catches fire whilst they are training, and the others try to tell him that he's on fire. However, because he is so used to their jokes and wind ups, he just says, "yeah yeah" and ignores them until the flames get so high that he finally realises for himself and is seen dancing about trying to put himself out. Well.................................

After about ½ an hour of high intensitive firefighting, someone turned around and saw Bullet with his back to the pump, in a world of his own, just watching us working. What he didn't know was that the heat from the exhaust of the pump had dried out the grass bank it was sitting on and had eventually caught fire. This fire was growing and had surrounded the pump.

"BULLET YOU'RE ON FIRE!" we shouted at him.

"Yeah, yeah, you concentrate on your job and leave me to mine" he sarcastically replied.

"NO BULLET YOU REALLY ARE ON FIRE!" we shouted back.

"Yeah, yeah" he replied with a disregarding wave of his hand towards us.

But then he turned around, away from us and towards the pump.

"OH SHIT!" we heard him shout out and then he began a stamping dance around the pump that anyone in Riverdance would be proud of.

We all, including John, stopped firefighting for a minute and just doubled up with raucous laughter as we watched Bullet's performance around the pump.

When he had finally stamped out the last bit of burning grass, he turned to us and, with an embarrassed look on his face, he too burst out laughing.

It was a good job that the owner of the boat house wasn't there at the time, as we finally regained our composure and returned to the task in hand, that of putting out the boat house fire.

We saved a beautiful wooden Thames Slipper Launch from the dock beneath the boathouse and only lost about 20% of the actual wooden building. We were rather pleased with that.

On return to station, after a couple of hours, Bullet's penance for his inattention was to make the tea for all of us, which was usually the sprog's job. As we sat around drinking our tea some stood up and attempted to imitate Bullet's dance moves to much laughter. But it was all done, in Kenny Everett's words, "In the best possible taste" and it was taken that way by Bullet, as well.

CHAPTER 19
New Career False Start

Every weekend we would take it in turns for 2 of us to go into the station and do a general clean of the engines. One November weekend in 1986 it was mine and Grahams turn. To be honest there wasn't much to do as we always kept the engines clean, but it was the running of the pumps and other equipment we had to check and sign up. This usually took about an hour.

This weekend we had completed all the checks and signed the pay book for an hours pay each and then decided that as we both weren't in a rush to have a cuppa. While Graham was making it, I took a wander to the notice board and took down the routine notices.

As I've previously said, these notices came out weekly and basically were the day-to-day items, news and goings on in the brigade. If your name appeared in them, it usually cost you cakes on station. Any excuse for a cake, like when I passed my BA course then a couple of weeks later, my HGV. That cost me a lot of cakes that month!

I was flicking through this week's notices when I came across a job vacancy.

"Fire Control Operator in the brigade control room Kidlington, expressions of interest to be submitted in writing by the end of the month to Fire Control Officer Ann Sadler".

Hmm I thought, Fire Control. I had by this time applied to 3 brigades to become a full-time fireman and got through to final interview with Surrey and Berkshire and had to pull out of the Oxfordshire process through illness.

We washed up our mugs and tided the sink. By now it was lunchtime, and the bar upstairs was open, so we went upstairs for a pint and the obligatory game of Sunday cribbage.

When John came in, I collared him for a quick chat and asked him about fire control and what he thought of it. He held the "girls" as he called them, in very high esteem. There were 2 men in control, Phil and John, but the other 16 were all female. Fire Control was looking to be a rather attractive option. It was fulltime, uniformed, shift work 2 days 2 nights and 4 days off, what the hell I'll give it a go I thought.

When I got home, I raided mums best writing paper and envelopes and wrote a letter in my best neatest handwriting to "express my interest". I posted it, first class, the same day as the deadline was only a week away. Then to be honest I completely forgot about it.

The run up to Christmas in the shop was manic with all the festive stock arriving and me having to find a place to store it in the tiny cellars was a nightmare. Being within the celebrity belt of London we often had stars coming into the shop. I once without knowing it served Ossie and Sharon Osbourne, only being made aware of this after they had left.

It brought back memories of living at the lock. My dad had a shrill whistle and whenever he made the sound, I knew there was a celebrity on a boat in the lock. I would come running out with my autograph book and dad would point to the boat concerned. I got many famous people this way, to name a few:

Ernie Wise

Kenneth Williams

Liberace, he even drew a piano by his signature

Linda Bellingham

Peter Davidson

Liza Goddard

The list goes on.

One drill night in December, John called me into the office, "There's a letter here for you from HQ" he said.

"I'm so sorry about the state of the envelope, I promise you I haven't opened it"

At the time the brigade was already starting to recycle things, which is fine, but when you get a letter in a used envelope that had been resealed by stapling it together, it was taking it a bit too far, and lots of stations were complaining about it.

I prised the staples apart and in it was a typed letter from Fire Control Officer Sadler. In it she thanked me for my expression of interest in the fire control operator vacancy but due to a change of circumstances the vacancy was no longer available.

She then said that my name and details would be kept on a list for any other vacancies that may arise. Oh well at least it wasn't a rejection I suppose. I folded the letter back into the now unusable envelope and went back to the drill night.

Christmas came and I was looking forward to a nice rest on the day, following a hectic couple of weeks in the shop. Mum made a nice breakfast of eggs and bacon, crispy streaky bacon just the way I like it. We then sat down and opened our presents.

It got to mid-morning and mum was prepping the Christmas dinner for us when the alerter went off.

"You're not on call surely? You did last Christmas day and got called out"

That famous chimney fire just as I put my knife and fork into the turkey.

"Yes, I am mum, same as last year. Those with kids get

Christmas Day off and the others get Boxing Day off. At least you haven't started cooking yet" I shouted as I ran out slamming the front door behind me.

When I got to the station the lads were still getting their kit on and someone called out to me,

"It's St Marys Church."

Blimey that's the big one in the town down by the bridge. That would explain why they were still getting their fire kit on in the station and not on the engine on the way to the fire. The church was less than a minute from the station and it is bad form to arrive at a job and still be getting dressed,

I jumped on the second engine and in the blink of an eye we were pulling up opposite my shop by the church.

The Christmas congregation were all stood outside the main doors shivering. Some were staying to hopefully go back inside, whilst others had done their duty and were on their way home.

As we walked up to the door, the vicar greeted us with rambling apologies and said,

"We were at full volume singing Oh Come All Ye Faithful", when the organist called out to me saying that smoke was coming out of the organ! It's an historical musical machine, please save it won't you".

John calmed him down and called for a couple of extinguishers to be brought off the engines. We went down the aisle to the magnificent organ with dozens of huge pipes reaching high into the roof like mighty stalagmites.

There was a faint whiff and haze of smoke in the air. I took off my helmet and as the smallest on the crew that day, started crawling under the keyboard and pedals. I quickly found the cause of the fire and called for a CO_2 extinguisher. The electric fan motor, that blew the air through the organ, had overheated, and was smouldering. It was this motor that has replaced the huge bellows, that in the good old days were pumped by small children during the services. A couple of squirts of the extinguisher and it was out.

With a bit more poking around with a screwdriver, I had it freed from its mountings and passed it back out to John.

"No more singing today I'm afraid vicar, but it could have been worse I suppose."

The Vicar laughed and after explaining the situation to the congregation he wished us a Merry Christmas and we headed back to the station.

Mum was glad to see me back so quickly and we shared a giggle when I told her what had happened.

The rest of the Christmas week went quietly with a few shouts but nothing much to mention.

When New Years Eve arrived, we had decided to have a party in the fire station social club. The parties up there were legendary. As I look back now, I see all that community spirit gone. There are no more social clubs and people are losing track of each other. This began with the smoking ban in all Oxfordshire County Council buildings, which had a serious knock on to the social clubs.

This night, though, was going to be a goodun. We had decided to go all in, and the blokes all hired black tie suits for the night, with the women going the whole hog and dressing up to the nines.

As with all these evenings we had to designate 8 of us to not drink, or in some cases cut right back, to keep the engines on the run.

The party was in full flow with Tony running the disco, and all the fancy lighting, and I had brought Anna up to the club for the first time. The poor girl was getting the 3rd degree off the other women, but in a nice way and all in good fun.

With about an hour or so to go before midnight the lights suddenly came on in the club, closely followed by the station sounders and our pagers going off.

The language and expletives from the 8 of us on the run was magnificent as we put down our drinks and or girls if we were dancing and headed down the stairs.

I went straight for a tally and got the driver of the first engine. John called out "Alarms at Phyliss Court."

Now let me explain about Phyliss Court. Henley-on-Thames is posh not doubt about that. But Phyliss Court is POSH!

It's a large site on the banks of the Thames made up of private apartments that line a long well-manicured driveway, that even in the 1980s would go for over a million each. There was a hotel, restaurant, croquet lawn, private moorings, a floating marque, the car park was full of Rolls Royce and Bentleys etc. you get the idea.

We set off down the road, the journey was only a minute, and I swung onto the drive, through the magnificent gates and headed on down between the neatly trimmed hedges, but I had forgotten, and could not see then in the dark, the speed humps along the route.

Bang! I hit the first one at quite a speed and the front of the engine bounced up and came down with a thump and screeching of suspension springs, but that was not all. At just that moment John had been putting his helmet on and as he took off from his seat his now helmeted head hit the roof of the cab with such force that the crest of the helmet dented the inside of the cab roof.

But what was even funnier was that his helmet had been pushed hard down on his head and he looked like a cartoon character trying to pull it off.

"DARRAN" he shouted.

"Sorry John" I replied trying my best to stifle a giggle when I saw him, but it didn't matter as the roar of laughter that came from the rear cab sort of rescued me.

I slowed right down and crept over the remaining road humps and pulled up at the magnificent front doors.

There was a manager and staff there to meet us. It appeared to be a false alarm caused by cosmetic smoke from a dancing show they were having, which had set off the alarms. John and some of the others went into the ballroom just to make sure and when they came out, they were all laughing their heads off again. They couldn't still be laughing at John, who had managed to unstick his head from his helmet before he got off the engine.

"What's so funny?" I asked as they climbed back onto the engine.

"All the old dears were really impressed with us as we walked into the ballroom. They said we really shouldn't have gone to the bother of dressing up for them" John told me.

"Eh?" I replied then it dawned on me they could all see our

bow ties and bright white shirts under our fire kit! We certainly were the smartest fire crew of the night.

CHAPTER 20
Disobeying an Order

1987 New year, new job. After the Christmas rush in the china shop, I decided I needed to find another job while I was still trying for a career in the fire service. Fortunately, Tony's sister Pat worked in a materials importer very close to my home. They had a vacancy and were very supportive of the fire service, so in January I started in the goods in dept. of Makowers. The one disadvantage was I now missed my regular trips to the post office to see Anna.

Now Makowers was in a lovely orange brick and flint-built warehouse, which was typical of the old Chiltern buildings and was only 1 min from home, which meant a longer time in bed, tee hee. Sadly, the building is gone now and replaced with very dull timber frame affordable housing. The only difference between them and the china shop, was that when I got a fire call, they stopped paying me, but the shop hadn't. I wasn't too worried as the fire pay was in fact more than I was being stopped.

A strange thing happened to me there one day. There was a shout from out in the yard and when I looked out there was a post office van on fire. I know nowadays my reaction would be different, but then I ran away from it and onto my moped and started heading down to the fire station way before my alerter had sounded. Much older and wiser I would now just go back into work, grab a couple of extinguishers and put it out before the engine arrived. It's strange how experience and the loss of youth changes one's approach to things.

Makowers was one of the country's top importers of fine

Chinese silks. They really were something to behold with bright vibrant colours and patterns and were as light as a feather.

I would have to check each roll in and double check the length on the ticket on the roll with the delivery note. Just 1 yard out would cost easily £100s. One day I was quietly checking in some material when the alerter went off. As I ran past the admin office I called out "I'm off on a shout" and the girls in the office, one of which was Tony's wife, would make a note of the time for the pay deduction. My moped was parked out by the front door with my helmet hung up just inside to keep it from the rain etc.

I got down to the station and the lads called out, "wood mill on fire with cylinders involved near Ipsden." Ipsden was well off our ground and into Wallingford B11 station. This was going to be a long drive. As we were pulling out of the station the chatter in the back was of a questioning nature. We were wondering if we were going to be first in attendance as there was no mention on the turnout of a make-up. This is where the first engines in attendance need more help and "make-up" for more engines, and this would normally be on the turnout printout, but it wasn't. Then we started wondering what sort of cylinders were involved? If they were acetylene, then we were in trouble. Because of the makeup inside the cylinder and how the liquid was stored, these cylinders were extremely volatile and explosive when damaged or heated, but again no mention on the printout, just "cylinders involved."

As we were heading down the Fair Mile out of Henley fire control came over the radio.

"Bravo Six Lima on your way to this incident you will come across Bravo One One Lima, they have been involved in an RTA.

There are other appliances mobilised to them you are NOT TO STOP but to proceed to the fire incident. All received? OVER."

John answered "All received HI Bravo Six Lima over"

"Bravo Six Lima roger, M2HI to standby," control replied.

(HI was the callsign for our fire control, or rather the abbreviation for it. As I was to learn, and in fact teach it later, the full callsign was M2HI. The M2 was the home office designation for a fire service, the H was the telecommunications area we were in, Hannington, and the I was for Oxfordshire. It was in the regulations that in all radio transmissions the full callsign had to be said once, so fire control always said it at the end, but usually so fast it all merged in to one mutter.

"So that explains a lot" said John turning around to address the 4 of us in the back and being met with astonished faces.

The next 10 minutes or so were then taken up with debating what had happened to them, and where exactly they were, we soon found out.

We were all leaning through the gap between the rear and front cabs, (we didn't have seat belts in the rear in those days), looking for any signs of an accident, when as we rounded a blind right hand bend with a steep grass bank on the right, in a wooded lane, there was a Renault car in a ditch to the left, and then as we went further around the bend "SHIT, BUGGER, BLOODY HELL AND WHAT THE F***" were some of the expletives from us as we were met with the rear of Bravo One One Lima on its side across the road. The ladders had fallen off and were laying in the ditch just past the Renault. But there was no sign of the crew.

Tony was driving and he looked across at John as he was slowing down to a crawl.

"Yes Tony we are stopping!" John said without having to be asked.

"I don't care what control told us"

As we slowly pulled past the roof of the engine we caught sight of the crew, they were sitting on the grass bank in front of the engine, all smoking like chimneys.

We stopped just past the engine, and all jumped out.

"Is everyone ok?" John asked

"Yeah, just shaken up a bit" Kev Fenemore their Sub O replied.

"We didn't stand a chance, she came flying round the bend on our side of the road. We swerved to miss her but hit on the corner and flipped over up the bank" He explained.

"Is she ok?" John asked

"Yeah, we've all be very lucky" Kev replied.

"Look we were told not to stop and there are others on their way to you. If you are sure you are all ok?" John went on.

"Yeah, honestly we're fine you go on John" Kev replied

"OK well look after yourselves and I'll call you later. Come on you lot back on the truck" John said.

They gave us the thumbs up as we drove off.

A mile or two further down the wooded road we started to see a large volume of smoke rising from in the woods. As we came into a small clearing, we could see a large wooden workshop well alight and slap bang in the middle was the cylinder, and yes it was an acetylene cylinder!

We pulled up in the gateway about 50 yards from the fire and set about running out the hose from the engine. Now with this cylinder involved in the fire, but not actually on fire, we were not allowed close to it, so we had to set up ground monitors.

This involved getting the roof ladders off the engines and placing them upside down on the ground, so the hooked end was on the soil and the ladder then formed a shallow ramp. We then wove the hose in and out of the rounds of the ladder and with the branch (nozzle) attached we tied it down firmly.

We now carefully move the ladders to point towards the fire and the cylinder. The water was slowly applied through the hose and as the resulting jets slowly raised up into the air we made adjustments to the ladders, so the water was falling where we wanted, mainly to land on the cylinder and cool it down. Once this was done, we could retreat to a safe distance and just watch.

The fire in the woodworking shed was almost out after an hour, with just hot spots still smoking, but we couldn't go anywhere near as the cylinder was still in there.

Every 30 minutes we would turn the water off and observe the cylinder from a distance. If the water on the surface of the cylinder evaporated quickly then the cylinder was still hot and too dangerous to approach, so water was applied again and checked 30 minutes later, and so on until. it remained wet when the water was turned off.

Over 30 years later we still do this evaporation test, but we also have Thermal Image Cameras (TICs) which when pointed at the cylinder give a heat picture and temperature readings to help us monitor more accurately the cylinders.

The Scandinavians have an even more drastic way of dealing with acetylene cylinders involved in fires. They bring in snipers who put a bullet through the cylinder and therefore release the pressure from a distance safely.

Now Divisional Officer (DO) Collins had arrived at the incident and John had briefed him as to what the situation was. We were now looking into how we could leave this scene safe.

We found a metal 8-yard skip approx. 50 yards from the fire and in the open. John told us to fill it with water to see if it had any leaks, which we did, and it amazingly didn't.

So, the plan was to move the cylinder carefully and slowly to the skip and gently submerse it in the water to keep it cool and safe, simples!

Now in the backs of all our minds was Sub O Wixey, who the previous year had been killed when a much smaller acetylene cylinder exploded by him.

The DO told us to get a couple of jets charged with water and be ready to cover them, by them I mean him and John. They were going to lift the cylinder themselves, and between them carry it the 50 yards to the skip and put it in it. This was extremely brave, and they wouldn't listen when we offered to do it. These two officers were ones who had the utmost respect from the brigade because they led by example, and this was no different.

So, with us on the two charged branches the two of them slowly approached the cylinder in the smouldering workshop. John Gosby carefully put his hand on the cylinder to feel the temperature, and after he gave us the thumbs up, we knew we were on for the move.

DO Collins took the neck end of the cylinder and our John, gently tipped it over to him and then lifted the bottom of the cylinder. Now these are not light items and as they slowly made their way to the skip, we could see that John was having difficulty maintaining his grip on the wet and muddy bottom of the cylinder, and about halfway there it slipped from his hands and landed with a loud thump on the ground. The look of sheer panic on the faces of the two men was mirrored by the ducking of everyone else around. There was a moment of silent anticipation of an explosion that thankfully didn't come.

John carefully picked up the cylinder again and the two officers resumed their heavily laden walk to the skip. When they got there, they heaved the cylinder over the rim of the skip and gently lowered it to the bottom under the water.

We all breathed a sigh of relief. As the officers walked over to us, they both had grins of relief on their faces and we heard the DO say to John,

"Don't you ever do that to me again. I need a clean pair of trousers now" and they both laughed.

With the main danger now removed we could finally get into the workshop and put out the remaining embers. We found out that the fire had been caused by sparks from some welding the owner was doing which landed, unseen, in a pile of sawdust. By the time it was noticed, it was too big a fire for him to deal with.

When we had finished putting the fire out, and had made up the kit, the DO had received a radio message letting us know that all the Wallingford crew had been checked by ambulance crews and were all fine and not going to hospital, which was great news.

We headed home a different way as we were told the road was blocked with the recovery vehicles for the car and Bravo One One Lima.

When we got back to station it was well past lunchtime and we had been out for 5 hours and were getting very hungry. Graham and Tony decided to go to the local Wimpy for a late lunch, and as Makowers weren't paying me I thought I'd join them. I was starving, and I did love the special sauce Wimpy had in their ¼ pounder.

About an hour later I walked back into Makowers and poked my head in the office to tell the girls I was back, and quickly told them what we had been to, as they were always keen to know.

The following drill night John called me into the office for a quick word.

"Darran when we get back off a call, can you please go straight back to work. Someone saw you in the Wimpy last week and told your bosses who asked me what was going on. I know it was innocent and I explained it to them, but obviously, someone has a bee in their bonnet"

"I don't believe it" I replied. "We were starving, and it's not as if it cost them anything as they stop my pay when I'm on a shout, and as we had booked off from the station, I wasn't even being paid by the fire service!"

"I know I know Darran" John said in his calming manner. "As I said I've smoothed it over, but just a word to the wise, OK?"

"Yeah, thanks John."

It was a shame, but I never trusted anyone in Makowers again, and I thought they were my friends. Maybe it was simply a case of jealousy I'll never know.

As a note, the Fire engine Bravo One One Lima, to our amazement was rebuilt by the insurance company, and the Wallingford crews swore blind that it never drove in a straight line again!

CHAPTER 21
New Beginnings

Drill nights came and went, fire calls came and went, winter plodded into spring into summer. One Sunday afternoon after the bar had closed, I went downstairs in the station and had a mooch around. I'll have a look at the routine orders, I thought, as I really hadn't looked at them for a few weeks. John usually read out the important stuff on drill nights, but he was on holiday now, so they had been, not missed, but overlooked.

I pushed back the large bulldog style clip on the board and released a month's worth of the orders.

I sat on the platform in the muster bay and flicked through the earliest one first, 5 weeks old.

It had the usual notices of changes in operational procedures, promotions, people who had just passed courses, that will cost them cakes, I thought. Then on the last page were listed the vacancies.

Most were for departmental transfers, so of no interest to me, but right at the bottom was a listing for vacancies in Fire Control. Oh, great I should receive a letter soon I thought. I read on and saw that there was a closing date for applications of the 25th of August. That was last week!

Why hadn't I received a letter as I was promised?

Had I missed the boat?

I copied the details down and a telephone number for Fire Control Officer Ann Sadler, I'll call her first thing Monday.

I skimmed through the other orders and put them back on the wall under the bulldog clip and headed home.

Makowers was on summer shutdown, so I had a week off work. When 9 o'clock came the following morning, I was straight on the phone to HQ.

A lady answered the phone, and I asked to speak to Fire Control Officer Sadler. I was put on hold for a while then a click and a voice answered.

"Hello Ann Sadler, how can I help"

"Hello, good morning, my name is Darran Gough, and I'm retained at B6 Henley" I started.

"Hello Darran, what can I do for you?" she replied.

"I've only just seen the advert in RO's (Routing Orders) for the vacancies in fire control, and was wondering if it was too late to apply, please?"

"I'm afraid you've missed the closing date Darran" Ann Replied.

"Arh but" I countered, "I applied last year and was told that the vacancy no longer existed, but my name would be kept on a list, and I would be contacted the next time a vacancy arose"

"Yes, I remember the vacancy last year, the person withdrew her resignation. But we don't keep lists I'm afraid" she said.

"But I have a letter stating that I would be put on a list and contacted, signed by the Chief Fire Officer"

There was a slight pause then,

"Oh really. I'll need to investigate that. Can I have your phone number and I'll call you back as soon as I can Darran"

I passed on my details and thanked her for her help.

That Wednesday evening at drill, the station phone rang and a few minutes later John, who was back off holiday, called me into the office.

What have I done now? I thought as I climbed the three steps to his office door.

"It's Fire Control for you" John said handing me the receiver.

"Hello" I said.

"Hello Darran, It's FCO Sadler," said the voice on the other end.

"That phone call we had on Monday has really opened a can of worms" she said. "I had completely forgotten about the letters we sent out and I've had to look back and contact many more in your situation"

"OK" I answered.

"So basically, you were missed and shouldn't have been. I will put an invite in the post tomorrow inviting you to a testing and selection evening next week at Head Quarters"

"Oh, OK, thank you very much" I answered.

"In the invite" she continued, "will be some material about the brigade that you will need to learn as you will be tested on. There will also be a typing test, a visit around fire control, and if you pass you will be invited for an interview. But it will all be explained in the letter. Is that ok?"

"Yes, thank you very much" I replied.

"Good then that's that. Can you pass me back to your Station Officer please?"

"Certainly, thank you again," and I passed the receiver back to John and he nodded to me to leave, which I did.

As I walked back into the muster bay, I was greeted by Nick,

"What do Fire Control want you for then?"

"Something and nothing" I replied, knowing damn well that this reply would wind Nick up something chronic. Nick always had to know the gossip, and I knew John wouldn't tell him anything.

I could see his mind working overtime wondering why Fire

Control would be contacting a lowly fireman in Henley. I left him looking frustrated, and me with a wicked smirk.

He would find out sooner or later, but it would be fun making him sweat for a bit.

A couple of days later a padded envelope arrived at home from HQ, inviting me to a "Fire Control Vacancy Selection evening."

It was at HQ in Kidlington next Tuesday starting at 1900.

There would be several tests including,

Typing both dictation and text.

Memory test involving the information sheets enclosed.

A visit to the Fire Control room.

If the tests were successful, then an interview would also be carried out that night.

Wow, all to be completed in one night, that's going to be heavy I thought.

In the envelope was a leaflet on Oxfordshire Fire Service and the county of Oxfordshire with various statistics on population, number of emergency calls etc.

There was also a list of stations with what appliances were stationed there and their callsigns.

These must be the memory test, so for the next few days I was reading them through and through, with mum testing me on them.

I was fine with these, but what worried me was the typing test. I was not, and still am not, a touch typist. I call myself a very fast two fingers and thumb merchant. Also, my spelling is not great, as my proof-readers of this book will agree with. In those days spell check didn't exist, and I knew that speed and accuracy would be paramount in that job.

I gave John a ring and asked him if I could book the next Tuesday afternoon and evening off call. I knew he would say yes but it was manners to ask.

John was still the only one on the station who knew what was going on.

"Nick is going nuts trying to find out what you are doing Darran" John told me.

"I knew he would, he'll find out if I get the job, but for now he can sweat a little more" I replied.

"Anyway, good luck Darran, just keep your cool and listen to the questions carefully. You'll be fine I'm sure of it."

Now I was in a bit of a conundrum. "Should I go in civies or uniform?" I asked John.

"Either would be fine" he replied.

Now I made a decision, that I would make again when I went for full time ops in West Midlands, 21 years later. It was a selection evening, and I wanted to stand out from the crowd, so I went in uniform.

My Boss at Makowers gave me the afternoon off as well. He also promised to keep it quiet as there were so many fire station relatives working there. He knew when I started there that it would only be for a short while, whilst I pursued a career in the fire service, and he was very supportive towards me.

Tuesday lunchtime came and I left Makowers and went home for a shower and some food. Mid-afternoon I borrowed mum's car, as to go all that way to Kidlington on my moped would have been daft to say the least.

I put on my uniform which mum had ironed for me and set off via the station to book off call for the rest of the day.

Mum's Austin Metro was a good little runner, and I made it to

HQ in an hour. I parked in the CO-OP car park opposite the building and sat and re-read the leaflets again, as I had an hour to kill.

The HQ building was made of two floors and was long and thin. The offices were to the right of the entrance, and the fire station was to the left. I knew from reading the leaflet that it had a fire engine A6L and the brigade's rescue vehicle A6R. The station also had some demountable pods, a couple of which were the mobile control unit A06X, and the support pod (Canteen) A06S. Both of which I had seen on incidents, the canteen one introducing me to the delights of corned beef and tomato sauce sandwiches.

I didn't know it at the time, but the fire station end had an extension build on top, and this was where Fire Control was.

1845 came and I thought it was time to leave the security of the car and venture to the reception. I had seen a couple going in, and now as I wasn't going to be the first, I made my move.

I was welcomed at the door, by a uniformed lady, and after taking my name I was shown into the conference room. Here were a few people milling around and lots of tables set out much like a school exam room each having a name on the table. I found mine near the back of the room.

Looking around I got the shock of my life because I recognised someone, Chris Hirons. Chris was retained in Banbury and had been on my basic course at Didcot last year. We greeted each other with the same stupid question,

"What are you doing here?"

We laughed and chatted for a bit.

Chris said he had thought about wearing uniform but had decided against it. Have I made a blunder I thought?

Spot on 1900, another uniformed lady called for our

attention and asked us to find our tables and take a seat, there were twelve of us in total. She introduced herself as Fire Control Officer Ann Sadler. Arh nice to put a face to the voice, I thought.

She introduced the others in the room explaining their role for tonight. There were 3 fire control staff, Divisional Officer Oliver, I knew him as the brigade training officer, and a couple of civilian admin staff.

On four of the tables were computers and keyboards, mine being one of them. During the evening the twelve of us would be split into groups of four to rotate through the various tests.

"Each test would be strictly timed, as we had a lot to get through tonight," she explained.

"Those sat at computers are to remain here, the next four tables are to go upstairs with Senior Fire Control Operator Brown, and the final four are to go into the next room" FCO Sadler instructed us.

I sat staring at the keyboard. It had been a while since I had used a computer and the keys were the stand-up clunky type, not the flat ones as on laptops nowadays.

When the other eight had left the room, we were given our instructions.

Firstly, we would be given a copy of a text from a fire service manual. In ten minutes, we would have to copy as much as possible, accurately into the computer.

Then we would have a dictation test. Some simulated fire calls would be played on a tape to us, and we had to record as much information as possible.

Ok so here we go then.

When told to, I turned the sheet of paper over and found the instructions to pitch a 13.5m ladder from the drill book.

I started typing away but slowly as there was no spell checker in those days and I wanted to be as accurate as possible as I knew that the dictation test could be a problem.

The ten minutes flew past and when time was called, I had finished the passage, and was re-reading it for any errors, so reasonably happy there.

Someone came around to each computer and printed out our work for marking.

We were then given clear screens again and got ready for the dictation test.

The tape started and a female voice came on saying there was a fire at, and gave an address, said what was on fire, there were people inside, and what road the address was off of.

It was very calm and controlled.

There was a pause and then another voice started with another fire call. This went on for 5 – 6 minutes, and I think we got through 5 calls in that time.

Again, someone came to print out the calls for marking. I was reasonably happy again. I had to guess some spellings of the road names but overall, I thought I had done ok.

We then sat waiting for one of the groups to return. When they did, we were moved into the memory test.

It felt like a school test this time. In front of us was a blank answer sheet, which we had to put our names on. There were 30 questions to answer, they were all single word answers.

We turned the question paper over and started.

Some were asking names of stations.

Some were asking the type of crewing at a station.

Some were asking the call sign of stations. The main station being Kidlington. So off I went reciting my list.

A06L WTL Water Tender Ladder

A06R Rescue Tender

A06X Control Unit

A06B Box Van

A06F Flat Bed

A06S Support unit (Canteen)

It also had two tractor units to pick up the four demountable units.

The only one I struggled with, and I did for a long time, was remembering A11 Burford and A12 Bampton, and which one was which.

When time was called, I was very happy with my efforts, being quite sure I had answered all the questions correctly.

This time we didn't have to wait for a group to arrive. We were taken upstairs to see the Fire Control room. We entered through a remotely controlled door and immediately up a narrow flight of stairs, which sounded a bit hollow. They creaked like plywood steps. At the top of the stairs was a door on the left. The person leading us knocked on the door and went in telling up to wait in the corridor.

When the previous group came out, we went in.

The room was about the size of a reasonable living room. To

our left immediately inside was a smaller room which was the Day Duties Office. At the far end were two glass doors with two printers behind them. These were constantly printing out any computer activity the operators did on the system, which was why they were behind the soundproof doors.

Immediately in front of us was the supervisor's desk. This was on its own. On the desk were two computer screens, one for the supervisor to work on, and the other to monitor the work of the other two operators.

To the right of the screen was a sloping desk with lots of coloured buttons and a dialling pad. This was where the incoming emergency calls came in and the radio messages were made. The supervisor could press in a button and listen to the call that an operator was taking.

Behind the supervisor was a large metal trolly. This had about a thousand small flip cards in it and was a throwback to the old mobilising system. These cards, alphabetically, held every street in Oxford City as well as every village and town in the county. On the cards was the PDA, Pre-Determined-Attendance. This was a list of the first 15-20 fire engines to that location, nearest first. The computer system now displayed this, but the old cards were still hanging on as a safety belt.

Later, I would witness this trolly being knocked over just like the game of 52 card pick up, only a thousand times worse!

Behind the supervisor, on the wall, was a white board with Red Blue and Green magnetic holders each with a name in. Alongside these were written in black pen their current location in the brigade. I recognised a few of the names, especially the blue ones, as they were the B Division officers, the ones I met the most on incidents.

Along the wall to our right was a long desk with two identical operator positions on it. Behind them on the wall was a huge Ordinance Survey map covered by a clear Perspex sheet. Drilled through the Perspex were holes at every fire station. In the holes were coloured plugs with the callsigns of the fire engine on. I quickly looked at Kidlington, and to my relief I had got every callsign right.

Next to the map was another white board with handwritten addresses on. Alongside the addresses were some of the plugs from the map board. This indicated the current incidents, and the fire engines assigned to them.

We were introduced to the supervisor, Brenda, and the operator in the room, Gill. The third person on duty, Jane, was out on her meal break. To this day, now 36 years later, I am still friends with them, and we have the occasional meet up over lunch at a garden centre.

Brenda began to briefly explain all that went on in the room but was interrupted by a loud and grinding buzzing noise. Gill punched a flashing light on the switchboard and said "Fire Service"

Brenda then pushed the identically lit button on her switchboard and began listening to the call.

We all listened intently, me more than the other three as I recognised the address as one in my own station ground. Brenda called out

"Bravo Six Lima, Bravo Six Whisky and Bravo Golf One Zero."

The two Henley fire engines and the nearest officer.

Blast, I thought, I've missed a shout.

It was all over in less than a minute and looked so smooth and professional.

Brenda continued talking to us, while Gill phoned the officer to tell him of the incident.

A few minutes later I heard John call up on the radio booking mobile to the incident. It seemed strange hearing him from this end.

We couldn't wait to hear what else went on as we were ushered back downstairs. Anyway, I would hear from the lads when I got back, what I had missed.

We returned to the conference room and sat down.

A few minutes later FCO Sadler came in and said,

"I'm now going to call some names out. Those that I do, will you please come with me.

My name was the fourth to be called out. I stood up and moved to the door. Was this a good thing or a bad thing? We would find out soon enough.

Of the six of us called out were 3 women and 3 men including Chris Hirons. The FCO came out and took us out of sight and upstairs again into another office.

Once the door was closed, she turned around and smiling said,

"Congratulations you are the top six from the tests this evening. If you wait here, you will, one at a time, be called into an office down the corridor for an interview."

With that she left the room and we all turned around to each other and there were smiles and puffs of relief all around.

A couple of minutes later a lady opened the door and called my name out, I was first up.

I was taken down the hallway and into a small office. Sat there behind a table were FCO Sadler, DO Oliver and a lady from HR.

I was directed to a chair, but before I sat down, I made a

point of shaking everyone's hand, something I had always done when going for a job.

"Well, Darran, what do you think of the process this evening so far?" DO Oliver began.

"It's been a bit stressful to be honest, but so far I've really enjoyed it sir" I replied.

"Good" he replied.

"You have come tonight in your uniform, can you please explain why when a dress code wasn't on the letter?"

"Yes sir, well as I'm in the brigade, and applying for a post in the brigade, and I'm proud to wear the uniform, I thought I ought to wear it sir"

"That's a fine answer thank you"

FCO Sadler then asked, "you do understand that this is an office-based role and not going out on the fireground?"

"Yes, I'm fully aware of the role I'm applying for. I have applied for whole time three times with different brigades and been unsuccessful. I figure maybe that role is not for me. I do like being in a supportive role. I work at a local theatre doing the lighting and special effects, which I enjoy more than being on stage and I see the fire control role as a very important and similar supportive role."

The DO then asked, "we have no retained members in the control staff. Do you intend to keep on your retained duties at Henley if you are successful?"

"Yes sir, during my four days off I do intend to stay retained at Henley, and I don't see that being a problem."

All three looked at each other and nodded.

The questioning went on for a further ten minutes or so and I was then asked if I had any further questions.

To which I just asked, "when will we find out the results of the evening?"

"By the end of the week" came the reply.

I was then thanked for attending and told that I could now leave for Henley. I thanked them all and shook their hands again before I left. I didn't see the others, as I was shown down a different staircase to the front exit. Driving home I began re-running the interview over and over again, wondering if I had answered the best I could.

I got home and over a cuppa, told mum all about my evening.

I tried to put it out of my mind for the rest of the week, but on Friday I went home for my lunch to be met by mum at the front door holding a large white envelope.

It had Oxfordshire Fire Service franking on it. I carefully opened it and "YES, YES, YES," I shouted at the top of my voice, I was in.

The start date was four weeks away in mid-October, but I had done it!

I went down the fire station club that evening, and John was already there. I asked him for a quick word in his office.

This drew some looks from the likes of Nick, Rick and Tony, but they would find out soon enough.

I started by saying, "John, I need to put in a change of cover please."

He rolled his eyes as this was always an issue when people wanted to change their cover, especially as I was a valuable full cover fireman.

"Why?" was his tentative response.

"Because I've got the job in fire control," I blurted out.

"Oh congratulations" he said with a huge smile on his face. "When do you start?"

"About four weeks' time."

"Darran that's fantastic news, are you going to let the cat out of the bag now?" meaning Nick and the rest.

"Yes, and I think the first round is on me tonight."

We headed back up to the bar, obviously both smiling.

"What the hell is going on?" Nick asked, as we walked back in. I told everyone what had happened over the week.

My whole life was about to go through a huge change, and I was so excited by it. It seems strange writing this at this particular time in my career, 38 years later.

I am one year from retirement.

My wife, Sheila, is a Watch Commander in Northampton Fire Control.

CHAPTER 22

Back to School...... AGAIN

The following few weeks went by in a blur. I went shopping in WH Smiths for folders, pens, pencils, highlighters and really anything else I thought I might need.

Although my hair wasn't really long, I decided to have a cut and look really smart for my first day, so I booked a haircut for Saturday afternoon at Rudi Kartel's, my usual hair salon in Duke Street, ready for the Monday morning.

When I went in the usual girl greeted me and after showing me to the chair, she started with the clippers first. I explained that I was starting a new job on Monday. She was halfway through cutting my hair, and we were chatting away when........ BEEP BEEEEEEP BEEP BEEEEEEEEP the pager went off for a fire call. Now she was used to this as it wasn't the first time I had been called out there. She whipped the gown off and as I ran out the door I called back,

"See you in a bit to finish it."

The run from Kartel's to the station was a short one and I got the drivers tally for the first engine B06L. I then swore big time as I saw the turnout ticket.

Factory fire Watlington, Make pumps 6

Bugger! we were going to be out a long time, and I wasn't sure we would be back before Kartel's closed as it was already gone 2pm.

John Gosby drove into the yard, and I just called out "Make pumps 6 Watlington both" John nodded as he ran past the engine to collect his kit.

I got the engine started and the radio turned on as the rest of the crew started arriving.

As John climbed into the cab he said, "We've only got eight on call, so as soon as we've got four we'll go." This meant I didn't have any option I had to go, I couldn't stand down to finish getting my hair cut.

Rick and Tony climbed in the back, and we pulled out the station. As usual I stamped on the floor button to put the two tones on as we approached the police station at the bottom of West Street. I carefully went through the STOP sign and down Falaise Square, named after the town in northern France, that Henley was twinned with.

I swung left into Bell Street, still with the two tones sounding, bringing back memories of me as a little kid with mum racing out of the shops to watch what I was now driving, a big red fire engine, every little boy's dream in those days.

As we left Henley John asked "Darran do you know where we are going?"

"Not really" I replied. "I know it's in Watlington, but exactly where I'm not sure."

"OK well we've got 15 mins or so before we get there, Tony here's the map book, see if you can find it" John said as he passed the book to the back for Tony.

As Tony leaned forward for it, he glanced at me and let out a laugh.

"Darran what the hell do you look like?"

"What do you mean" I replied.

"Have you gone half punk or what?"

Then it dawned on me, the girl was using the clippers when the pager had gone off and I had a nice step right in the middle of my head where she had got to and stopped.

"You're joking" I said.

"Nah I'm not eh John?"

I glanced over to see John with a large grin on his face, oh great I thought. I'd better get back in time or I'm going to look really stupid on Monday.

We climbed up the Chiltern Hills towards Nettlebed and turned towards Watlington running along the top of the hills. Then I had to slow to decent down Howe Hill into the Oxford Plain. Going down was fine, but we used to call this hill, "How the hell do you get up it?" As it was really steep coming the other way, so I had to ride the brakes trying to avoid brake fade, where the brakes get too warm and begin to fail with the heat, as we descended it. Now the hill was in a forest and as we came out of the trees at the bottom, Tony said "I can't find it on the map John, but I guess it's that way" pointing at a huge plume of smoke in the distance.

That's it I thought I'm not getting back before closing time now.

As we pulled in, we were met by B09L (Watlington) and B11L (Wallingford) fire engines. We were directed to the rear of the industrial estate where there was a small brook running along the edge. I was to set into the river and start pumping from there to the fire as the fire hydrants weren't that strong with water pressure.

Tony helped me with the heavy work of connecting the large suction hoses together and humping them into the brook. Rick went off shadowing John, no surprise there then.

As we were doing this Henley's secondf engine B06W arrived closely followed by the two engines from Thame B07L B07W. So, there's the six pumps asked for.

I don't remember much of the incident after that as I was the pump operator for my engine, The usual job for the driver was to operate the pump upon arrival at an incident, but we were all trained in operating the pump in case the driver was needed elsewhere.

The fire was brought fairly quickly under control and then the rumblings and rumours of relief crews started. As Henley and Thame were two pump stations, they would have had a standby crew from another station moved into them to maintain fire cover. Henley's usual was from Goring and Thame's was from Wheatley.

It made sense to release an engine from each station to return and let the standbys either go home or come onto the incident as relief crews. That was the case this afternoon. My engine and the first from Thame were released to go home. We left the fireground around 1615.

So, 20 minutes or so drive back to Henley another 20 minutes to tidy up and I just might get back to the salon to get my hair cut finished.

Approaching Howe Hill, I had my foot to the floor. Usually, you needed two or three gear changes going up the very steep hill. I hit the bottom of the hill in fifth gear and as the gradient increased, I could feel the engine slowing. I had to "Double Clutch" to change down a gear, as explained in the previous chapter on my driving course. When there is no synchromesh in the gear box to maintain the speed of the gears, when you want to change down a gear you depress the clutch, pull the gearstick back into neutral, release the clutch pedal, rev the accelerator the while the revs are still high, depress the clutch pedal again and push the gear stick into the lower gear. It sounds complicated, but with practice I could do it in under a second. If it took any longer the engine would slow down too much and then you might stall or have to start from a standstill, which on Howe Hill was not a good idea, with 400 gallons of water on board.

So I went from fifth gear to fourth to third to second slowing all the time and finally crawled over the crest of the hill doing no more than 5mph.

This journey home was going to take longer than 20 minutes.

As I drove back down the Chiltern hills and along the Fair Mile into Henley a sarcastic comment came from behind me along the lines of "Blimey I didn't know we had another shout," Rick had never passed his ordinary driving test, so I ignored him.

By the time we got back to the station, John told me to just hang my kit up and go, I ran back down into town but being a Saturday, they had closed at 4 O'clock not the usual 5 O'clock, so all my rushing was for nothing.

I was going to look a right numpty on Monday morning now with half a haircut.

I had an idea, I ran to Boots and caught them just as they were closing, I pleaded my case and got in to get a tube of hair gel. I thought at least I could flatten the long half of my hair to match the short side for the week.

Monday morning came, and with lightly gelled hair, so I didn't look like a Bril Cream boy I hit the road for Kidlington.

I parked again in the co-op car park, not really knowing where to park, and headed for reception.

I introduced myself at the front desk and was told to take a seat. Whilst sat there Chris Hirons walked in and we just gave an acknowledging smile at each other, and once he had introduced himself at reception, he joined me, and we chatted until a third man joined us. His name was Greg Cole age 17 and he had worked in the brigade stores.

Spot on 0900 someone came walking down the stairs and introduced herself. Jackie Brown was a Senior Fire Control Operator and the current Fire Control training officer. She would be taking us through our first 4 weeks of basic training. Once we passed that we would be allocated to a watch, Red, White, Green or Blue.

She led us into a small conference room which was going to be our home for the next month. We asked where we should park our cars and she showed us the fire control spaces in the drill yard. Jackie said we should move them now or we would get a ticket from the Co-op otherwise. We dropped our bags and moved our cars there and then. After that we went across the drill yard, following our new teacher like good little boys to the stores, Gregs's old haunt, and as we entered, he got some ribbing from his old work mates. On the counter were the all too familiar large cardboard boxes containing yet another set of uniform, only this time with white shirts. We carried them back across the drill yard and back into the main building where all three of us disappeared into the gents to get changed.

We reappeared wearing bright white long sleeve shirts with the sleeves rolled up with that new starched look and folded creases in all the wrong places. We had a clip-on black tie, black heavy surge trousers and black slip-on shoes. We had small sized black epaulettes to slide onto our shoulders that said, "Oxfordshire Fire Service". All control rank markings were the same as the operational but about two thirds the size.

We also had a peaked cap and brigade badge to put on it. Here was another difference between operational and control. The badge was the same, but fire control had a red disk behind it whereas operational didn't. I still have mine to this day.

We also got a large black raincoat with FIRE embroidered on the left breast.

Left in the boxes were three more shirts and another pair of trousers, and we were also given black socks. They were 100% nylon, and they were the same awful ones I had been issued when I joined the retained in Henley, I only ever wore them that first day. They were heavy and boy did they make my feet sweat and smell.

By this time, it was tea break. We were shown up in the control suite where the kitchen was and where Jackie had supplied us trainees with tea and coffee and where our milk was in the fridge. We were to make our drinks in the kitchen but had to take them down to the HQ mess to drink them, leaving the rest room in the control free for the on-duty staff, so as not to disturb their breaks.

After tea we returned to the classroom. Jackie then went through the program for the next four weeks. The first week was really getting to know the brigade. The stations, which we had already learnt, the officers and their callsigns, and the beginning of learning the mobilising procedures in theory. What gets mobilised to what, when and why.

The second week we would start to use the computerised

mobilising system, in the training room. We would learn our way around the system and how to mobilise fire engines. We would also start to take simulated fire calls. We would be taught how to ask open and not closed questions. This means how to ask a question that does not give the caller an answer. In a panicking state a person will agree to anything you say, for example.

Questioning about a car fire.

Closed question

Operator. "There's no one in the car is there?"

Caller "NO" without even looking.

Closed question, you are telling the caller the answer, not good.

Open question

Operator. "Is there anyone in the car?"

Caller "No" but they will have had to think about the question and look, hopefully.

Week three the mobilising would get more complicated with the introduction of major road, railway and waterway mobilising.

Week four would be more practical work finishing off with a theory paper and a practical assessment in the control room in front of the Fire Control Officer Ann Sadler.

However, at the end of each week we would have a smaller test paper so Jackie could assess how we were getting on.

At the end of the fourth week, if we were successful, we would be posted onto a watch for another four weeks following their shift pattern. We would not be counted as a full member but would continue to be trained up by the watch officer and would be taking most of the emergency calls.

At the end of the four weeks, we would be practically assessed by Ann Sadler again but also by the Divisional Officer in charge of brigade training DO Howland. If successful we would then be counted on the full watch strength. We would be on probation for a year and following another assessment at the end of that time we would fully qualified Fire Control Operators and get a nice pay rise as well.

It all sounded a bit daunting, but exciting at the same time.

The next couple of weeks were a bit of a blur.

Learning the county topography was a key element as was special mobilisation areas like the M40. At the time the M40 only went as far as Jn7 for Oxford. But the extension was being built to link up with the M42 in Birmingham. Not only did we have to learn the current motorway, but also the new one with its new junctions and the current construction access points as well.

We began to start playing with the computer mobilising system. The screens were bulky, and the display was green writing on a black background. We had to learn what information went where in the "New Incident Field" whilst asking questions of the caller, played by Jackie, and thinking of the next question.

This is a skill that can't really be taught, but it got me into a lot of trouble at home and still does today. I can be reading the paper, listening to the TV and the wife or whoever is talking to me all at the same time. I'll then be accused of not listening to them or not paying attention, so I'll repeat the last couple of things they have said, with a nonchalant look on my face.

The first two weeks of training and the exams at the ends of the weeks went well for me, passing with high 80% both weeks. Chris and Greg however were beginning to struggle a little and either failed or only scrapped through.

As week three started all three of us were summoned up to the Chief Fire Officer, Maurce Johnson's, office. We had no idea why but were told it was nothing bad. We checked our uniform, and all smartly walked down the long central corridor through HQ and waited in his reception room with his PA until we were called in.

Maurce Johnson was a "fireman's fireman." Like all officers he had come up through the ranks, usually from several different brigades on the way. I discovered later that he was a very clever politician with a very sharp mind.

He welcomed us into his office, and we had some general chit chat about how we were finding things etc.

He then got onto the real reason for inviting us in. He commented that we were the first retained firemen, me and Chris, (not Greg), to be employed in the fire control. He then said something that initially didn't hit home with me at the time.

He said, "There might well be people in the brigade who think you don't belong there, and they may put some pressure on you in different ways. Please don't let them, and if it does occur then report it immediately."

He then wished us well in our training and bade us goodbye. As we walked back up the corridor, we all looked slightly puzzled and discussed what had just happened with Jackie when we got back to the classroom, she didn't say much, but if she understood what the Chief meant then she didn't let on.

Week three whizzed past with more and more practical call taking and mini exercises. There were times when one at a time we were allowed to work in the control room. Jackie would send me into the room and, under the supervision of the watch officer, be allowed to take radio messages from the crews and enter them onto the incident logs. If it got busy or an assistance message came in, I would quickly slide my chair back and let the watch operator take over.

It was during one of these sessions that I was watching over Phil Baskerville's shoulder when I saw something that has stuck with me to this day.

When people are on the radio, they take on a different persona, similar to the phone really. They feel that they must sound official, sometimes coming over as pompous in my opinion.

There was one officer who whenever I heard him on the radio, later in my career, I always thought, "Here we go chapter 1 verse 3 paragraph 7." He always seemed to feel why use one word when ten would do and sound better.

Jackie had been drumming it into us that we must enter exactly what the person on the other end of the radio said.

It was him on the radio now, when Phil was taking an informative message for an incident he was at.

He started,

"One private motor vehicle in close proximity to a semi-detached domestic brick-built property"

Phil typed. One car near to a house

I absolutely loved this and left the room with a massive grin on my face.

More and more now I was feeling that I was being separated from Chris and Greg and I didn't know why. Was I doing something wrong? I had to know.

End of week three test came and this time I hit the mid 90%. I didn't know what the others got, but at tea break I collared Jackie and asked her why I was being separated all the time. She quietly said that she had no worries with me or my performance whatsoever. She was happy for me to work on my own or with the duty watch in the control room. She had far more concerns about Chris and Greg and was beginning to believe that they would not pass the course, so was having to put in much more time and effort in with them than me. This made me feel really good about myself but also a bit guilty and sad for the others.

Week four, and I'm spending more time on my own in the control room with the duty watch while Chris and Greg are getting extra work in the classroom with Jackie. The duty watches are great and let me take everything except real call just yet.

To give them some free time from babysitting me I slope off to an empty office and do some revision.

This seems to be the pattern for most of the week, then Thursday arrives, test day!

We are taken into the classroom, which has had the tables laid out for the exam. We are given our papers and told we have one hour to complete them, and off we go.

I complete my paper in forty-five minutes but don't want to leave so I start again from page one and carefully re read every question. I'm glad I did as I pick up some silly errors and have almost finished re-reading it, when Jackie calls time. We all breathe a sigh of relief, hand our papers to Jackie and leave for a well-deserved cuppa.

Chris and Greg start going over the questions and what answers they gave, I hate doing this, I have ever since school days, it makes me start to doubt myself. After an hour or so Jackie comes to find us and calls me out of the room, I am going first into the practical assessment, leaving the other two in the rest room.

I enter the control room and take my seat at the far end out of the way of the duty watch who will be watching my performance whilst taking the real jobs. FCO Sadler is in the room as my assessor and Jackie will be in the room next door putting in the exercise calls on a dedicated line.

The FCO explains what is going to happen and it should take thirty minutes or so. The screen I'm using is switched into training mode so I can't turn out a station for real. I put my head set on, I'm asked if I'm ready, I nod back and the line buzzes with my first call.

I must admit it's all a bit of a blur but just over 30 minutes later FCO Sadler says

"OK Darran that's it all finished. Can you please put all the appliances you've mobilised back to their home stations, and we will clear the incidents."

After I've done this, I'm told to go downstairs to the classroom and wait there whilst the others have their assessment.

I unplug my headset and go downstairs via the kitchen for a cuppa and a breather, hopefully that's me all done.

About an hour later Greg enters the room after having done his test, and this time I can't help talking about what we've just been through. His whole demeanour is of a nervous person who feels that they haven't done too well.

Another hour passes and Chris turns up. Again, we discuss the calls as we all had the same incidents. Then Jackie sticks her head into the room and tells us all to go and have some lunch and to be back here for 1400. We all head out to the local baguette shop which is also the butchers. We wander down to a park and sit on a bench still dissecting the calls we had just taken.

1400 and we are back in the classroom, nervously awaiting our results when Jackie comes in and takes me out of the room and up to FCO Sadlers office. We wait outside until we are called in and told to sit down.

"Well Darran, how do you think that went?" asks the FCO.

"I think ok. I know I could have been a little faster mobilising a couple of the calls but......." then she interrupts me.

"Well, I think you've done exceptionally well. You've scored 95% on your written and 88% on your practical which is really pleasing, well done."

"Wow thank you" I reply

"Now" she continues, "We want you to go home now, and you are being posted onto Green Watch who start on Saturday, so you have tomorrow off then start to follow their shift pattern, OK?"

"Yes, yes that's fine" I say, but I was a bit confused as Green Watch was the only fully staffed watch with 4, Brenda, Jane, Gill and Angela, so I would be over staffing them, but I was happy with the posting.

"Now there is one other thing" the FCO continued, Greg and Chris have not passed and will have to do extra training and another assessment. That's why we've called you in first. Can you please just collect your things and leave without talking to them as we know they will not be happy". I acknowledged this and could gather my things and slip down the top floor corridor missing the classroom completely.

"Is there anything you would like to ask Darran?"

"No thank you I'm happy with that"

"Well done, Darran," Jackie added. "You nailed that today, I'm very pleased for you."

With that I quietly slipped out of the control suit, grabbed my stuff and left HQ without passing the classroom window. Now this was well before mobile phones, and I could have found a phone box to call mum, but I wanted to see her face so I had to wait for the whole drive home before I could tell her my good news. She was over the moon for me, and we had a special treat that night of a takeaway to celebrate.

I phoned John to let him know my news and that I would be following Green Watch pattern so I would be down for drill that Wednesday night.

The following evening, I went down to the Fire station social club and told everyone else my news, but I didn't stop long as I had an early start in the morning to get to Kidlington for 0800 start and I really didn't want to be late on my first watch day.

CHAPTER 23
Heart Breaking Decision

Saturday morning and up early at 0600, left home by 0630, arrived at HQ 0730 ready for the start of my first shift in Fire Control with Green Watch at 0800.

The day shift was 0800-1800 and the night shift was 1800-0800, two days followed by two nights then four days off.

I put my coat in my locker, took my headset out and walked into the control room. The room was a bit crowded as the hand over was just taking place. Reds were just going on their four rota days off. Teresa, the Senior Fire Control Operator, SFCOp, equivalent of an operational Sub Officer rank, was handing over to Brenda, the Green Watch SFCOp. She was telling Brenda all the incidents that were still ongoing, any crewing problems and generally anything of importance. All the Reds said hello and well done to me as they left to go home.

Once we had the room to ourselves, Brenda welcomed me to the watch. One of the first things she said was that she was surprised I was joining them as they were fully staffed. I agreed.

So important things first! Once the handover was complete, Gill took me into the kitchen to show me where to put my food and even more importantly as the newbie, how to make the tea and coffee and who had what.

Now drinks were not officially allowed in the control room, but on nights and weekends when HQ was empty rules were bent, so when I brought them in everybody automatically opened the top drawer in the units under the desks and in there were coasters waiting for their mugs. If any officer pressed the intercom on the door at the bottom of the stairs, everyone would have a few seconds to discreetly close their drawer. However, if it was a senior officer at the door who had a key, they may not use the intercom, so all we heard was the door "click" then you had 2 seconds or less to close your drawer as they climbed the stairs, and if you weren't sat at your desk at the time of the "click" a quick run around the desk or your neighbour would lean over and do the honours.

It all sounds a bit silly writing this now but at the time it was deadly serious.

The girls on Green Watch are amazing and I'm still in touch with them all, except Angela who has sadly passed away. Twenty six years later they all came to my marriage vow renewals to celebrate 20 years with my wife, who also worked in Oxfordshire control, but much later. She is now a Watch Manager in a different county fire control as Oxfordshire has now closed and merged with Buckinghamshire and Royal Berkshire to become Thames Valley Fire Control. It was fantastic to have the old gang back together.

Brenda was the Senior Fire Control Operator, (SFCOp) and in charge of the watch. She had served in the RAF before joining the fire service. She has a strong personality, and I didn't experience many officers arguing or disagreeing with her, but I mean this in a nice way. She was a sound leader, who knew her job inside out, and I was very happy to have her as my first SFCOp.

Jane was the Leading Fire Control Operator, (LFCOp) or second in charge. Jane had just come back off maternity leave.

Gill was one of the Fire Control Operators, (FCOp)

Angela was the other FCOp

Angela was single but the other 3 all had partners in the brigade.

One abiding memory of that first day was Angela. She was in her late 50s and had the appearance of being a bit fierce. Her hair was always tied up in a tight bun, and she had very long fingernails which tapped when she used the keyboard. She had been in Fire control for over 20 years and was very experienced but had not wanted to be an officer.

Brenda told me to sit at the far console and I could start by taking the radio and admin calls during the crews shift change.

By mid-morning Brenda casually called out to me

"Darran the next emergency line is yours."

"Yikes" I thought. I didn't have long to wait.

"BUZZ BUZZ BUZZ BUZZ" The first red button with 999 on the AD9000, which was basically the switchboard, burst into light,

I pressed it in, at the same time calling up the new incident format on the computer screen"

"Hello Fire Service" I answered.

"This is Aldershot connecting you to 01295 765438" The BT operator announced.

I repeated the number as I entered it onto by screen with 9AL at the start.

I then said again "Hello fire service, what's the problem?"

"There's been a car crash under Bodicote Flyover and people are hurt" came the reply from the caller

I went to the incident type field to type in RTA persons trapped. (Road Traffic Accident). In modern terms it's now called an RTC, (Road Traffic Collision), allegedly because there is no such thing as an accident, someone is always at fault.

Before I could start typing Angela, who was stood behind me and listening into the call, was leaning over me and tapping the screen to show me where to type. I knew this as I had been doing this for a few weeks now and it was a bit off putting.

I continued with the call asking for the nearest town which was Banbury, and the road which was the Oxford Road.

Every time I moved my cursor Angerla was leaning over and tapping the screen. I was beginning to get a little flustered now.

I hit the PDA key. PDA is the Pre-Determined Attendance, this is a list of the nearest, in this case 10 fire engines, to the address. I selected A01L - WT and A01W - R the two Banbury engines. The WT meant the station light and bells would be operated as it was a wholetime station. The R was for the second engine as it was retained, like I was in Henley, so their alerters would operate as well.

I then selected A06R WT which was the heavy rescue engine stationed right beneath where I was sitting.

I checked all was correct and then hit the red send button on my keyboard twice. As I did, I heard the sounders operating outside, alerting the crew downstairs.

I told the caller we were on our way and cleared the line.

While I was doing all this, Gill had been on the phone to the ambulance requesting their attendance.

I now called the police and requested their attendance.

Brenda had already mobilised the nearest officer so all that was left for me to do was to go to the white board on the wall behind me and write up the incident address and pull the 2 plugs from Banbury and 1 from Kidlington from the county map and put them in the holes alongside the address.

"Well Darran that's your first call then, well done," said Brenda.

I just smiled, still in a bit of a state of shock at how quick everything had happened, the sign of a good team I thought.

I didn't know it at the time, but a circle of fate had started. This exact address and incident would reappear 21 years later. When I would be leaving Fire Control to become a full-time firefighter in the West Midlands. As OIC of the retained crew in Banbury I would be attending the same type of incident and location as my last call before leaving.

One thing was bothering me though, Angela. It was really off putting her leaning over me and tapping the screen all through the call. I knew it was with the best intentions, but it wasn't helping. When we had a quiet minute, I mentioned this to Gill. She had seen what was happening and agreed with me. She quietly mentioned it to Brenda who I believe had a word with Angela and it didn't happen again.

At lunch time I went out to the rest room and got my sandwiches out of the fridge and sat down contemplating the morning. Then lots of laughter came through the window from the yard below. The rescue tender was back from the RTA, and I looked out to see two fully grown men, (I do remember their names!) running around the yard wearing ladies' bras, panties stockings and suspenders! What the hell, I thought, then one of the audience, laughing, shouted up to me as I was leaning out of the window.

"It's for Children in Need, they lost a bet."

And with that I threw a fiver out the window for them.

The afternoon went much the same I took more fire calls, without tapping, and enjoyed the day.

1730 ish and I had tidied the kitchen and emptied the bins as White Watch began to arrive. Brenda told me I could go as I didn't need a relief.

Sunday was very much like Saturday, it's nice and quiet with only the weekend routines to carry out but then came my first ever night shift.

I arrived for duty at 1730 ready for 1800 start. When I entered the room Greg was sat in my chair. He had stayed late to get more practice in before retaking his exams. Brenda asked if I minded sitting out for a bit longer so Greg could do more practice. Of course I didn't mind.

Once Blue Watch had left, I checked if it was OK, and I slipped out to make the tea. There were only 3 of Green Watch in tonight, Angela had the night off.

Greg stayed for about an hour or so and then making his thank yous, he left. I didn't want to ask him how he was doing, it felt a bit rude, but I thought it was the better route rather than rubbing it in that I passed, and he hadn't.

Brenda explained that the normal night routines were to start lunch breaks from 1930 and hour each then from 2230 until midnight we relaxed. From midnight to 0630 the night was split in 3 so each got a couple of hours out in the rest room sleeping.

As I wasn't counted yet she told me to go into the locker room at midnight and I could stay there until 0630. I liked this in one respect but in another I really wanted to get into the routines.

When an emergency call came in a buzzer would sound just outside the rest room to tell anyone on break that a call was in progress. Usually they didn't do anything, but if in a couple of seconds, a second buzzer sounded then they would return to the control room as the two in the room would both be taking calls.

However, if the person was needed urgently back then there was an eject bell. This was a manual button on the SFCOs desk. This was a loud bell and if that sounded you ran back.

I decided that if I heard a buzzer go off, I would return just out of sheer nosiness. Also, I wanted to make sure I would react when needed. I did have a habit of missing the retained alerter for fire calls when in a deep sleep so much so that I had an amplifier box made.

The evening passed quietly, all meals eaten undisturbed. Come 2330 I went out and set up my little camp bed in the admin storeroom and got my sleeping bag out. Officially we weren't supposed to sleep so we all had our own means that could be stored in our own locker. Years later we were officially allowed to have a sleep so as a control we bought two z beds which was much better.

About 0200 the buzzer went off and I jumped up with a start and then quietly stuck my head in the control room door to see Gill taking a call. I thought I'd make myself useful and looking over Gill's shoulder I saw the address and started writing it on the board with the fire engine plugs.

After the call Brenda told me I really didn't need to respond, but I politely said that I would like to, to get in the rhythm of things.

I then went back to my camp bed and sleeping bag.

I was woken at 0630 by Brenda knocking on the door. I put all my bedding away and made the tea.

By 0730 Blue Watch were starting to arrive and I was told I could go home. Even though I had slept through the night it wasn't a good sleep and I was tired driving back to Henely. I got home around 0845 had a cup of tea with mum and then went to bed. Now that was a deep sleep, with no alerter to worry about I was out for the count.

I woke around 1500 feeling much refreshed. Mum asked if I

wanted a meal before going back to work, but Jane had said she would bring some food in that evening, so I just had some toast.

Walking back into the control room that evening Chris was sat in my seat this time. Again, I stood back and made the teas. Jane had brought in dinner for us, and it smelt fantastic and was filling the kitchen with a lovely aroma of a steak and kidney pudding.

Chris went to leave about 1900 but he collared me for a chat.

He congratulated me on passing. He said that Jackie had given them a bit of a rough time after I had left. He had only just scrapped a pass on the paper but failed miserably on the practical. They were doing practical after practical all day and then had another test at the end of the week. If they didn't pass it, that was it, they were out.

Wow, I thought that seemed a bit rough, but I suppose they know what they are doing.

Chris left and I went back into the control room. As it was Tuesday night it was busy as half the brigade retained stations had their training session, the other half on Wednesday night. We had more than the usual appliance movements with stations meeting up for training or travelling to Rewley Road for heat and smoke work.

I took most of the radio work that evening and after 2100 When we knew the HQ building was empty and the retained downstairs had gone home, Jane went out and brought back into the control room 4 heavily loaded plates of steak and kidney pudding, veg and mash. We all ate in silence, primarily enjoying the food, but also keeping an ear out in case we had missed anyone in the building. The food was wonderful but so much, I couldn't finish it.

I volunteered to do the washing up. Whilst I was out in the kitchen, I heard a scream and then laughter coming from the control room. I ran back in to see what was going on. Gill was by console one, with Jane sat at the desk but with her right leg up on the desk. She was crying out in pain, with Gill and Brenda in absolute fits of laughter. As I entered the room Gill turned around with what I now know to be an Epilady hair shaver. She had been demonstrating it on Jane. As I entered Jane said,

"Fancy trying it Darran?"

"Umm no you're alright I give it a miss thank you" I replied

But with that Jane swung around off her seat and Gill made a lunge for me. I just swerved out of the way and tried to get back out of the door. By now Jane and Gill were in hot pursuit and I had to run down the corridor and lock myself in the toilet to escape the evil hair pulling machine.

After five minutes or so I crept back out and finished the washing up. When I thought it was safe, and with a tray full of cups I gingerly returned to the control room to looks of disappointment and ridicule.

"You wimp," Gill greeted me with.

"I'll have you know I'm very religious," I retorted.

"I'm a devout coward".

They all giggled as I handed out the mugs of tea.

I did have a restless night's sleep later, listening out for the door handle and a sign that the Epilady was coming back.

The next four days off were taken up with getting back on the run with the retained. We had a few shouts during the time, and it seemed strange that I now recognised the voices on the other end of the radio now.

First day back on the Sunday we were greeted by the news

from the off going watch that Greg and Chris had failed their assessments again and had been let go. This came as a bit of a shock to me, and I was sad that the two I had spent the last month with weren't here anymore.

Chris later actually went on to get into the wholetime operational crews. I heard he had a difficult time in training school again and sadly only lasted a year or so on ops before taking his own life. I didn't hear what Greg did, but he didn't go back to stores.

So, this left just me on an overstaffed watch, I wonder what was going to happen to me now.

I continued the next couple of weeks taking emergency calls, answering the radio and admin lines, Basically Gill and Jane had it easy as I was doing almost everything when in the room, and I was loving it.

The control Christmas dinner was on the horizon. It was booked in a pub in Kidlington, and we were all given a menu list to choose from. A starter, main, and desert. The day after was to be my second practical assessment and if I was successful, I would be counted a full watch member.

The assessors would be FCO Sadler and DO Howland, who was the brigade training manager at the time. DO Howland was also coming to the Christmas dinner with us. To save me travelling home, Angela offered me her spare room for the night as we would be between day shifts, which I gratefully accepted.

The meal was wonderful except for one minor hiccup. I can't remember who organised it, but they forgot to bring the menu choices with them, and someone had DO Howland's Satay starter and he got lumbered with French onion soup.

The following morning back in control, the list was found

and oh shit! I had had the DO's starter, and he was about to assess me, fan bloody tastic!

"Well Darran, you'll be starting with minus 5 points for each satay stick" he said as they gathered in the control room, with a wicked grin under his big black moustache.

"I am so, so sorry sir," I stammered.

"It's alright, I quite like French Onion soup anyway", he grinned.

The FCO explained I was to take the exercise as I had done with the previous one four weeks ago. The only difference was this time if a real emergency call came in, and I was in a position in the exercise where I could stop and take it then I should.

I nodded I was ready, and off I went. About halfway through the exercise a real call did come in and I seamlessly switched from training screen to operational screen and took the call.

After about forty-five minutes the FCO called a stop to the proceedings, and I could see both her and the DO scribbling on their clipboards.

I was told to go out for a cuppa, and I would be called back when they were ready.

I didn't need a drink, I just slumped down in the rest room chair and was grateful that was over.

After about 15 minutes the FCO popped her head around the door and asked me to go to her office. In there was DO Howland and Jackie waiting.

The FCO wasted no time.

"Darran that was an exceptional performance." she started.

"There were some minor errors which Jackie will go over with you later, but you have passed and are now a full watch member.

"Well done Darran" said DO Howland and Jackie in stereo, and they left the room.

"Right your watch posting Darran", the FCO continued.

"As you know Greens are overstaffed. I would like you after this tour is finished, to take six days off and then join Sylvia Ploughman on Blue watch on Boxing Day for two tours then in the new year, after two days off join Mags and White watch as your permanent watch."

"Oh right," I was a bit taken aback with the complexity of it all, especially over Christmas.

"As I said, and I'll say it again, well done Darran" the FCO finished with.

I explained the moves to Brenda, who like the others was rather confused by it all as was I.

But still I had passed and was a full member of fire control, on probation for a year, or rather ten months now.

The following night shift we decided to have a celebratory and leaving takeaway. Now the takeaway restaurants in Kidlington were well aware of our situation when we placed our orders. As we couldn't leave the building, they would have to deliver to us. The Chinese we used was within sight of the control room. We would place our order and get the price and delivery time. From the control room we would see the delivery man leave the restaurant and walk over to our side entrance. As we were on the first floor, above the appliance bay we would lower a rope down from the window, Rapunzel style. He would then tie the rope to the handles of the delivery bag, and we would haul it back up through the window. We would empty the bag, put our money in it, including a tip, and lower it back out the window to him. Pleasantries exchanged he would then untie the bag and walk off.

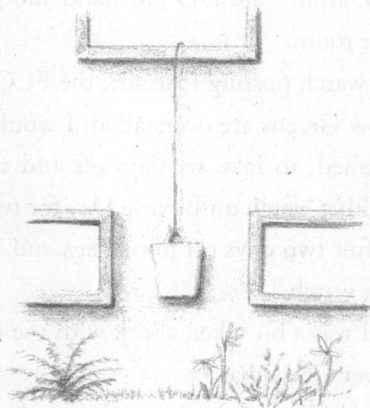

One night, much later and with a different watch, we got caught out. We had ordered a curry and had decided to use some small tables to sit around and eat rather than at our desks.

Now much like drinks, we weren't supposed to be eating in the control room. We were tucking away into the meal when the dreaded "click" of the control front door was heard. In sheer panic I picked up the small table and shuffled into the side office and slid the door shut, just in time to hear the main door open. I was listening hard and recognised the voice of DO Meadows. After saying hello to those still in the room, the aroma of the curry gave us away.

"Come on out whoever is hiding in there," he announced.

I sidled out leaving the tables behind me. Everyone burst out laughing and when we got the tables back out, he sat down and joined us.

I joined Blue watch on Boxing Day, Sylvia was the SFCOp Jean the LFCO and Sarah the only FCOp. The day was quiet being a Saturday as well. Jean wasn't on duty that day. Now even though I was a full member of the control I still wasn't supposed to be taking fire calls un-monitored yet. It got to the afternoon and Sarah was out for her break when an emergency line went. I answered it and started taking the call. I looked over to Sylvia to confirm I had the correct number of fire engines selected before I mobilised only to find to my horror, she was fast asleep! Hell, what do I do I thought.

I hit the eject bell to get Sarah back in as I couldn't shout over to Sylvia to wake up as I was still on the line to the caller.

Sarah came back in the door saw Sylvia, looked at me, I gave an expression of help me!. She came around behind me and saw the incident and the selection and quietly said "Yes go" so I hit the send button twice.

She then made a lot of noise plotting the incident on the wall board which woke Sylvia up.

She had a groggy look on her face which rapidly changed to embarrassment when she realised what had just happened.

Nothing more was said, but I wasn't happy with the situation I had found myself in.

The following day, Sunday, again was quiet and was beginning to drag when in the afternoon Sylvia got up with her mug to go and make a cuppa. As she approached the door the dreaded "click" happened again. In sheer panic again she shoved the mug up the back of her NATO jumper, a thick ribbed woollen jumper which we were issued with and was totally unsuitable for an indoor office environment.

As the main door opened, she backed away and wedged the mug between her back and the wall away from the entering officer, the Chief Fire Officer no less! He had popped in to say hello and merry Christmas to us. He stood in the doorway chatting for about fifteen minutes. I could see on Sylvia's face that she was becoming very uncomfortable with the mug digging into her spine.

The Chief was just summing up his chat and was aiming to leave, when the evil in me rose up. I'll teach her to fall asleep, I thought, so I asked him another question which I knew would necessitate a long answer in return.

Sylvia looked around at me with a face like thunder. After a further ten minutes chat the Chief bade us farewell. When we heard the bottom door click again, Sarah and I burst out laughing and Sylvia, who didn't really have a sense of humour, stomped out to the kitchen, fumbling to retrieve her mug from up her back as she went.

Now in the new year I moved yet again and joined White watch this time. Mags was the SFCO, Anita the LFCO and Andy the only FCOp. Now back into the working week with the festivities over, I had a visitation that brought the Chief's warning when we started back into focus.

I was accosted in the corridor of the control suite by the Chair of the Oxfordshire Fire Brigades Union. He introduced himself and came straight to the point.

"You will have to leave the retained" he said bluntly.

"Why?" I replied.

"Because you can't do retained and fire control."

"Why not?" I asked quarrelsomely.

"Because it's one man one job" he quoted back at me.

"But you let workshops do it. You even let the hydrant inspectors do it. What's the difference?" I argued.

"Look you will have to leave, that's all there is to it, or the union will have to take action. Those are the rules."

I was shocked, bewildered and angry. How dare this total stranger threaten me like that.

I told the watch what had just happened and they to be

302

honest, weren't sympathetic. Anita was the FBU control rep. I don't believe she had anything at all to do with what had just happened, but she wasn't surprised.

Next drill night I asked to see John in the office, and I told him what had happened. He wasn't at all surprised. He had been through the same. I hadn't realised this. John had been wholetime in Berkshire and during the strikes kept working retained. He faced the same sort of bullying and threats and left the wholetime.

I mulled this over for a couple of weeks. I now had a fulltime career that I absolutely loved. My part time job which I also absolutely loved was going to threaten my fulltime career.

I was not aware of any other union or representative body in the fire service although there was one other.

I discussed it with mum as well and her answer was the same as Johns. It had to be my decision.

I thought that as I was still in probation, I wanted to concentrate on passing that and didn't want any other distractions. So, with a very heavy heart I saw John the following drill night and completed a mis3 form and handed in my notice.

That night in the bar I bought everyone a drink and told them what had happened. I would still be coming down to the club though. I had to leave quickly as I could feel myself welling up and didn't want to make a fool of myself in front of them.

A week later I packed up my fire kit into the good old cardboard box and left it in the station for collection, and with a tear in my eye I turned off and handed in my alerter.

A couple of weeks later, on a dark winters Sunday evening, and I was walking downtown for a drink and a kebab. The roads were quiet as I walked down Greys Road towards Friday Street past the Meccano shop.

As I did, I heard one of the fire engines turning out and coming my way. I stood in the shadows as the blue lights bounced off the multitude of windows and watched as B06L flew past me heading towards Shiplake. I choked back a tear and waited in the shadows to see if the second engine was coming.

Sure enough in the distance I heard the familiar sound of the Bedford engine approaching. Remaining in the shadows B06W sped past, and I saw Graham getting dressed in the rear cab.

With that I turned around and went down Tuns Lane, a small lane behind the shop with no lighting and it was pitch black. I stayed down there for half an hour or more crying my heart out. It is still making me feel upset writing this 36 years later.

What gives someone the right, and how dare someone who doesn't know me, force or I am going to say it, bully me, to leave a job I absolutely love and put my heart and soul into and have spent years getting into.

Had I been weak buckling to their bullying?

Should I have stood up to them and told them where to go?

I didn't know what the consequences would be. I decided that I had done the sensible thing considering my position in probation.

But I also decided that night through my tears. I would never, EVER be bullied by anyone ever again.

The next book in this series *Fire From Both Sides* is currently smouldering and will ignite soon

ABOUT THE AUTHOR

Darran Gough was a child of the swinging '60s, born into a lock keeper's family. His childhood was idyllic, growing up at Shiplake Lock on the River Thames — he often boasts that he had a boat before he had a bicycle. However, a tragedy at the lock involving his father, Peter Gough, in 1982 would have a profound impact on his life and future career. What began as an innocent "I can do that" transformed into a lifelong career within the fire service.

Darran Gough's destiny was to be needed, but to be needed, someone had to face an unthinkable emergency. Whether in the direct heat of the action or coordinating from the control room, when a person made that dreaded call to 999, the adrenaline set in, and Darran flew into action. His unique experience spans all aspects of the fire service — from starting as a retained fireman in the heart of Henley-on-Thames, stepping behind the fire with Oxfordshire's control team while remaining on call at Banbury's busy two-pump station, to later transferring to full-time firefighting in the West Midlands.

Darran's comprehensive, 360-degree view of the fire service weaves together the lessons learned both on and off duty.

Muck, Grime and Sweat is the first installment of a firefighter's tale, as Darran Gough first steps into the uniform which will play a starring role in his life for the next 40 years.

www.ingramcontent.com/pod-product-compliance
Ingram Content Group UK Ltd.
Pitfield, Milton Keynes, MK11 3LW, UK
UKHW041038100825
461716UK00003B/135